Praise for Bjorn Lomborg's

Cool It

"*Cool It* is a highly valuable contribution to the climate-policy literature. In clear and concise prose, Lomborg diagnoses the problems plaguing contemporary climate policy, injecting a needed tonic of realism and common sense into the climate debate." —*National Review*

"A slim, brisk read. . . . [Lomborg's] timing couldn't be better." —*Men's Journal*

"Highly readable. . . . A plea for a rational discussion of social priorities." —*Foreign Affairs*

"Bjorn Lomborg's rational and compassionate suggestions would save more lives, preserve more wilderness and have a better chance of eventually halting man-made global warming than hysterical catastrophism, global treaties, and high-minded energy rationing. Read this ingenious book."
—Matt Ridley, author of *The Origins of Virtue*

"Lomborg affirms that the planet is warming, but questions why so much of the policy debate is framed around the idea of imminent catastrophe. This book dares to offer straightforward new thinking about how best to respond. Indispensable."
—Clive Crook, associate editor, *Financial Times*; senior editor, *The Atlantic Monthly*

"Brilliant! A devastating critique of the prevailing climate change hysteria. This book provides an overwhelming case for reassessing where exactly our policy priorities should lie if we are genuinely concerned with world welfare rather than with making noble—if futile—gestures that, at best, make us feel good but actually do a lot of harm."
—Wilfred Beckerman, Professor Emeritus, Oxford University

"At last we have a book that puts the hype of global warming into perspective. Bjorn Lomborg's eye-opening book, *Cool It*, examines and meticulously documents climate change's effects and proposed solutions. An extraordinarily timely and supremely useful book."
—John Naisbitt, author of *Megatrends*

Bjorn Lomborg

Cool It

Bjorn Lomborg is the author of *The Skeptical Environmentalist*. He was named one of the 100 most influential people in the world by *Time* magazine in 2004 and has written for numerous publications, including *The New York Times*, *The Wall Street Journal*, and *The Economist*. He is presently an adjunct professor at the Copenhagen Business School, and in 2004 he started the Copenhagen Consensus, a conference of top economists who come together to prioritize the best solutions for the world's greatest challenges.

www.lomborg.com

Cool It

The Skeptical Environmentalist's Guide to Global Warming

DISCARD

Bjorn Lomborg

Vintage Books

A Division of Random House, Inc.

New York

For Future Generations

FIRST VINTAGE BOOKS EDITION, AUGUST 2008

Copyright © 2007, 2008 by Bjorn Lomborg

The Library of Congress has cataloged
the Knopf edition as follows:
Lomborg, Bjorn.
Cool it : the skeptical environmentalist's guide to global
warming / by Bjorn Lomborg. —1st ed.
p.   cm.
Includes bibliographical references and index.
1. Global Warming—Government policy.   2. Climate change.
3. Environmental responsibility.   4. Economic
development—Environmental aspects.   5. Pollution—
Environmental aspects.   6. Globalization.
I. Title.
QC981.8.G56L657   2007
363.738'74—dc22   2007018646

Book design by Wesley Gott

Vintage ISBN: 978-0-307-38652-6

www.vintagebooks.com

12/08

Contents

Preface

Global warming has been portrayed recently as the greatest crisis in the history of civilization. As of this writing, stories on it occupy the front pages of *Time* and *Newsweek* and are featured prominently in countless media around the world. In the face of this level of unmitigated despair, it is perhaps surprising—and will by many be seen as inappropriate—to write a book that is basically optimistic about humanity's prospects.

That humanity has caused a substantial rise in atmospheric carbon-dioxide levels over the past centuries, thereby contributing to global warming, is beyond debate. What is debatable, however, is whether hysteria and headlong spending on extravagant CO_2-cutting programs at an unprecedented price is the only possible response. Such a course is especially debatable in a world where billions of people live in poverty, where millions die of curable diseases, and where these lives could be saved, societies strengthened, and environments improved at a fraction of the cost.

Global warming is a complex subject. No one—not Al Gore, not the world's leading scientists, and most of all not myself—claims to have all the knowledge and all the solutions. But we have to act on the best available data from both the natural and the social sciences. The title of this book has two meanings: the first and obvious one is that we have to set our minds and resources toward the most effective way to tackle long-term global warming. But the second refers to the current nature of the debate. At present, anyone who does not support the most radical solutions to global warming is deemed an outcast and is called irresponsible and is seen as possibly an evil puppet of the oil lobby. It is my contention that this is not the best way to frame a debate on so crucial an issue. I believe most participants in the debate have good and honorable intentions—we all want to work toward a better world. But to do so, we need to cool the rhetoric, allowing us to have a measured discussion about the best ways forward. Being smart about our future is the reason we have done so well in the past. We should not abandon our smarts now.

If we manage to stay cool, we will likely leave the twenty-first century with societies much stronger, without rampant death, suffering, and loss, and with nations much richer, with unimaginable opportunity in a cleaner, healthy environment.

Foreword to the Vintage Books Edition

This book received vastly different reviews upon its publication; *The National Review* said it is "a highly valuable contribution to the climate-policy literature" while *The Washington Post* called it a "stealth attack on humanity." Both of these viewpoints are representative of much of the commentary on my book.

Unfortunately, I think both viewpoints reprise the classical divide on global warming—either you believe it is an elaborate hoax or you think it is the unmitigated apocalypse. Yet both viewpoints are unsupported by the data. Indeed, we need to get our heads around the double facts that 1) global warming is both true and man-made, but 2) dramatic and fast CO_2 cuts are a poor way to deal with global warming and an extremely poor way to help the world and its inhabitants.

In *Cool It*, I try to stake out the sensible middle ground between global warming rejection and alarmism. Finding this middle ground is absolutely essential if we are to make a better future, both for people and the planet. This is why *The New York Times* described my book as part of the emerging "pragmatic center" on the global warming debate.

Most of the public debate seems to lack this center, and often it means that we end up with surprisingly inadequate positions, some of which I've been actively involved in exposing since the publication of *Cool It*.

One example has been the worldwide "lights-out" campaign from Sydney to Toronto, where environmental groups encourage entire cities to turn out their lights for an hour to emphasize the need for global warming action. Yet it seems nobody wanted to spoil the party by pointing that the event is immensely futile, underlines a horrible metaphor, and engenders extremely high pollution levels. Let's look at Denmark (the native country of the story of the "Emperor's New Clothes") for instance. The papers happily quoted WWF on how the event was a overwhelming success, yet the entire energy savings from the event (assuming people didn't use more energy later in the night to make up for lost time) was a full 10 tons of CO_2. This is equivalent of just a single Dane's annual emissions. Economically speaking, the effect of the entire collected good efforts of the Queen, the many participating companies, and Copenhagen city hall, as well as many other cities, managed to do only $20 worth of good. I'm sure this will make the future remember us fondly.

Moreover, what sort of message does turning out the lights send? As some conservative commentators like to point out, the environmental movement has indeed become the dark force, not metaphorically, but literally. Urging us to sit in darkness will indeed make us realize how utterly unlikely it is that we will be convinced to give up the advantages of fossil fuels. Indeed, it might make us realize how utterly dependent we are on fossil fuels. Curiously, nobody suggested that the lights-out campaign should also mean no air conditioning or heating, no telephones, Internet or movies, no hot food, warm coffee, or cold drinks—not to mention the loss of security when public street lights go out.

Ironically, the lights out also implies much greater energy inefficiency and dramatically higher levels of air pollution. Most of the people around the world when asked to extinguish the electrical lights, might turn to candlelights instead. It is cozy and seems oh so ecological. Yet, when measured on their light, candles are between ten and a thousand times less efficient than electrical lights—and this is compared to the maligned incandescent lightbulb, with energy-saving lights besting candles between a hundred and a hundred thousand times.

At the same time, candles create massive amounts of some of the most societally damaging pollution, namely particulate air pollution, which in the United States is estimated by the U.S. EPA to kill more than a 100,000 people each year. Yet, candles can easily create indoor air pollution levels ten to one hundred times the outdoor air pollution caused by cars, industry, and electricity production. Moreover, the whole family along with kids will conveniently be gathered around the coziness to fully inhale the extra fumes. Measured against the relative reduction in air pollution from the reduced fossil fuel energy production, candles increase the health damaging air pollution a thousand to ten thousandfold.

This seems more generally to be the state of much of our environmental debate. Because we're lacking the middle ground, we only hear the stories that fit the preconceived frameworks. You have undoubtedly read the story about a breakup of a massive glacier in the Antarctic, supposedly showing the ever-increasing effects of global warming. Yet we don't hear that the area was ice-free, possibly just some 400 years ago, without the help of global warming. We don't hear that the Wilkins glacier makes up less than 0.01% of Antarctica. And we don't hear the inconvenient fact that the Antarctic is experiencing record sea ice coverage since satellite measurements began.

While we all heard Al Gore talking about the dramatic hurricane years of 2004 and 2005, we've heard almost nothing about the complete absence of hurricane damage in 2006 and 2007. The insurance company Lloyds of London has now begun to fret that the absence of natural disasters is putting a squeeze on their premiums.

We are constantly presented with the stories underlining how temperatures are soaring, but over the past year, when temperatures worldwide have plummeted, we've seen the single fastest temperature change every recorded, either up or down. Yet, this rarely gets mentioned, although the stories abound. In January, Hong Kong was gripped with the second longest cold spell since 1885. Winter storms in central and southern China produced the worst winter weather in half a century. Snow fell on Baghdad for the first time in living memory.

Indeed, most of the public debate seems to lack a sensible climate this center. Instead almost all politicians in most nations are scuffling to promise ever stricter CO_2 cuts. This is evident also in the United States, where, as of this writing, all three major presidential candidates have set forward elaborate promises to deliver significant carbon reductions till by 2050. However, there is very little information on both the efficiency of such policies (how much less warming will we see?) and their costs (how many billions of dollars will it cost?). And for good reason. As you see on page 132 of this book, Al Gore 's proposal for a $140 carbon tax would hike gas prices by $1.25 per gallon, cut the U.S. emissions by half in 2015, yet have an almost immeasurable impact on temperatures—decreasing the average temperature in 2100 by 0.2°F. And the cost would be a dramatic $160 billion annually for the rest of the century.

While most U.S. politicians have only made promises but done little toward implementing them, it is perhaps worthwhile to look at politics in Europe where the extravagant promises are now beginning to be felt. A good example is the EU's newly instituted policy of cutting CO_2 emissions by 20% by 2020.

A 20% reduction in the EU, vigorously enforced throughout this century, would merely postpone global warming by two years at the end of the century. The temperature increase the world would expect to see by 2100 because of global warming would first take place in 2102. An immeasurable change. Yet the cost would be anything but immeasurable. The EU's own estimate for the cost is about €60 ($90) billion annually, which conveniently is down from its own previous estimate of almost twice that much. It is almost certainly a vast underestimate, since it requires the EU to make the reductions as smart as possible. As we shall see, the real price will likely become much higher. An economic cost-benefit analysis indicates that for every dollar we spend on such policies like the EU's, we only help the world about 30 cents.

However, the EU politicians are not content just talking about cutting emissions—they also want to decide *how*. They have decided that the renewable energy share in the union should be increased by 20% by 2020. This increase has no separate climate effect, since they have already promised to cut emissions by 20%, and because the extra spending will go to buying current technology not to produce much better future technologies. However, it does manage to make a poor policy decision dramatically worse.

Here, the debate in my native Denmark is instructive as the relevant government ministries have outlined what this decision will end up costing Denmark, and it gives a feel for the total cost for the EU. The costs for an increase in renewable of less than 20% (18 percentage points) and with less ambition (five years later by 2025) shows the total cost will be about $4 billion annually. And the benefit? If Denmark sticks to this decision throughout the rest of this century the end result will be to postpone global warming by the end of the century by five days. The average temperature that the globe would reach by Friday, January 1, we would postpone till the following Wednesday, after having spent more than $300 billion.

Is that a sensible decision? The total advantage into the infinite future from this Danish reduction to the world (measured on all relevant parameters such as lives saved, agricultural production increased, wetlands preserved, etc.) is about $15 million. Or for every dollar spent, we would do a bit less than half a cent worth of good for the world.

To put this into perspective, the $4 billion spending could double the number of hospitals in Denmark. And if we really wanted to do good, $3 billion could halve the number of malaria cases in the world. It would save 850 million lives over the century, avoid 250 million people getting infected every year. It would mean that the people of malaria-infected countries would live much better lives, become more productive and leave a world for the kids and grand-kids in 2100 that would be much, much better. The last billion could fund an eightfold increase in R&D of CO_2-reducing energy technologies, which in the long run would make it possible for everyone to reduce CO_2 much more dramatically, at a much lower cost.

So, what should it be? Achieve one the most remark-able improvements in history—reducing the frequency of malaria in this world by half—while dramatically increasing the possibility of solving global warming in the medium run? Or make a renewable pledge that looks good but does two thousand times less good and will change the global climate insignificantly?

And it gets worse. The price of $4 billion requires politicians to pick the cheapest way possible. Yet the politicians seem intent on often picking much more expensive solutions essentially doubling the cost or more. And the opposition—in trying to trump the government—insists that we should increase our ambitions to almost 40%. The cost would escalate to nearly $15 billion annually, with every dollar doing just a quarter of a cent worth of good for the world.

All of these considerations seem to be playing themselves out internationally. Using the Danish figures to extrapolate EU costs, the total cost is likely to be more than $225 billion annually, with every dollar doing just half a cent worth of good. And this is assuming that politicians pick the lowest-cost best options and that the opposition let's-do-more attitude doesn't win. I'm not sure this is clear. How do you measure what the best options are?

The same money could triple the global development aid budget. It could easily give clean drinking water, sanitation, education, and health care to every single human being on the planet, while tenfold increasing CO_2-reducing R&D.

And there is one final risk in dealing with global warming, we need to remember. If we persistently overworry and exaggerate the problems people will eventually tire of the entire discussion. There are already ominous signs that this is happening. A recent poll from National Center for Public Policy Research shows that while 18% of all Americans are willing to pay 50¢ on the prevention of global warmning, almost half are unwilling to spend even a penny more. A recent UK survey from the Environmental Transport Association showed that three in 10 people reckon there is too much publicity about global warming and more than half of these are "bored hearing about it."

Remember the bird flu? A couple of years ago, we worried extensively about a global influenza pandemic, which could kill hundreds of millions people. We worried mostly about birds in Europe and the United States infected with H5N1, while the risk lay mainly with the billions of birds living in close proximity to hundreds of millions in Southeast Asia. And eventually we got fed up with the scary news stories and moved on. This doesn't mean the risk wasn't there—if anything the risk of a global pandemic is probably increasing. But the fact that we overworried, focused wrongly, and then gave up shows a disconcerting analogy with global warming. We should have focused on smart policies to gain information, sequence diseases, and ramp up investments in research for vaccines, while avoiding the hyperbole that eventually drained interest.

Not worrying enough about global warming is as wrong as worrying too much. But if we keep sounding the alarmist doomsday drums on global warming, we are likely to enact ineffective and enormously costly policies, that will do little to help the world but much to drain us of our long-term stamina necessary for seeing us through. Instead of focusing on how intensely we should we be worrying, we need to worry correctly. We need to find the pragmatic middle ground and get smart, cost-effective policies to deal with global warming. But we will only get there, if we keep our cool.

It is my hope this book can help us get there.

Bjorn Lomborg
April 2008

Cool It

1

Polar Bears: Today's Canaries in the Coal Mine?

Countless politicians proclaim that global warming has emerged as the preeminent issue of our era. The European Union calls it "one of the most threatening issues that we are facing today." Former prime minister Tony Blair of the United Kingdom sees it as "the single most important issue." German chancellor Angela Merkel has vowed to make climate change the top priority within both the G8 and the European Union in 2007, and Italy's Romano Prodi sees climate change as the real threat to global peace. Presidential contenders from John McCain to Hillary Clinton express real concern over the issue. Several coalitions of states have set up regional climate-change initiatives, and in California Republican governor Arnold Schwarzenegger has helped push through legislation saying that global warming should be a top priority for the state. And of course, Al Gore has presented this message urgently in his lectures as well as in the book and Oscar-winning movie *An Inconvenient Truth*.

In March 2007, while I waited to give evidence to a cogressional hearing on climate change, I watched Gore put his case to the politicians. It was obvious to me that Gore is sincerely worried about the world's future. And he's not alone in worrying. A raft of book titles warn that we've reached a *Boiling Point* and will experience a *Climate Crash*. One is even telling us we will be the *Last Generation* because "nature will take her revenge for climate change." Pundits aiming to surpass one another even suggest that we face medieval-style impoverishment and societal collapse in just forty years if we don't make massive and draconian changes to the way we live.

Likewise, the media pound us with increasingly dramatic stories of our ever worsening climate. In 2006, *Time* did a special report on global warming, with the cover spelling out the scare story with repetitive austerity: "Be worried. Be *very* worried." The magazine told us that the climate is crashing, affecting us both globally by playing havoc with the biosphere and individually through such health effects as heatstrokes, asthma, and infectious diseases. The heartbreaking image on the cover was of a lone polar bear on a melting ice floe, searching in vain for the next piece of ice to jump to. *Time* told us that due to global warming bears "are starting to turn up drowned" and that at some point they will become extinct.

Padding across the ice, polar bears are beautiful animals. To Greenland—part of my own nation, Denmark—they are a symbol of pride. The loss of this animal would be a tragedy. But the real story of the polar bear is instructive. In many ways, this tale encapsulates the broader problem with the climate-change concern: once you look closely at the supporting data, the narrative falls apart.

Al Gore shows a picture similar to *Time*'s and tells us "a new scientific study shows that, for the first time, polar bears have been drowning in significant numbers." The World Wildlife Fund actually warns that polar bears might stop reproducing by 2012 and thus become functionally extinct in less than a decade. In their pithy statement, "polar bears will be consigned to history, something that our grandchildren can only read about in books." *The Independent* tells us that temperature increases "mean polar bears are wiped out in their Arctic homeland. The only place they can be seen is in a zoo."

Over the past few years, this story has cropped up many times, based first on a World Wildlife Fund report in 2002 and later on the Arctic Climate Impact Assessment from 2004. Both relied extensively on research published in 2001 by the Polar Bear Specialist Group of the World Conservation Union.

But what this group really told us was that of the twenty distinct subpopulations of polar bears, one or possibly two were declining in Baffin Bay; more than half were known to be stable; and two subpopulations were actually *increasing* around the Beaufort Sea. Moreover, it is reported that the global polar-bear population has *increased* dramatically over the past decades, from about five thousand members in the 1960s to twenty-five thousand today, through stricter hunting regulation. Contrary to what you might expect— and what was not pointed out in any of the recent stories— the two populations in decline come from areas where it has actually been getting colder over the past fifty years, whereas the two increasing populations reside in areas where it is getting warmer. Likewise, Al Gore's comment on drowning bears suggests an ongoing process getting ever worse. Actually, there was a single sighting of four dead bears the day after "an abrupt windstorm" in an area housing one of the increasing bear populations.

The best-studied polar-bear population lives on the western coast of Hudson Bay. That its population has declined 17 percent, from 1,200 in 1987 to under 950 in 2004, has gotten much press. Not mentioned, though, is that since 1981 the population had soared from just 500, thus eradicating any claim of a decline. Moreover, nowhere in the news coverage is it mentioned that 300 to 500 bears are shot each year, with 49 shot on average on the west coast of Hudson Bay. Even if we take the story of decline at face value, it means we have lost about 15 bears to global warming each year, whereas we have lost 49 each year to hunting.

In 2006, a polar-bear biologist from the Canadian government summed up the discrepancy between the data and the PR: "It is just silly to predict the demise of polar bears in 25 years based on media-assisted hysteria." With Canada home to two-thirds of the world's polar bears, global warming will affect them, but "really, there is no need to panic. Of the 13 populations of polar bears in Canada, 11 are stable or increasing in number. They are not going extinct, or even appear to be affected at present."

The polar-bear story teaches us three things. First, we hear **vastly exaggerated and emotional claims** that are simply not supported by data. Yes, it is likely that disappearing ice will make it harder for polar bears to continue their traditional foraging patterns and that they will increasingly take up a lifestyle similar to that of brown bears, from which they evolved. They may eventually decline, though dramatic declines seem unlikely. But over the past forty years, the population has increased dramatically and the populations are now stable. The ones going down are in areas that are getting *colder.* Yet we are told that global warming will make polar bears extinct, possibly within ten years, and that future kids will have to read about them in storybooks.

Second, polar bears are **not the only story.** While we hear only about the troubled species, it is also a fact that many species will do *better* with climate change. In general, the Arctic Climate Impact Assessment projects that the Arctic will experience *increasing* species richness and higher ecosystem productivity. It will have less polar desert and more forest. The assessment actually finds that higher temperatures mean more nesting birds and more butterflies. This doesn't make up for the polar bears, but we need to hear both parts of the story.

The third point is that **our worry makes us focus on the wrong solutions.** We are being told that the plight of the polar bear shows "the need for stricter curbs on greenhouse-gas emissions linked to global warming." Even if we accept the flawed idea of using the 1987 population of polar bears around Hudson Bay as a baseline, so that we lose 15 bears each year, what can we do? If we try helping them by cutting greenhouse gases, we can at the very best avoid 15 bears dying. We will later see that realistically we can do not even close to that much good—probably we can save about 0.06 bears per year. But 49 bears from the same population are getting shot each year, and this we can easily do something about. Thus, if we really want a stable population of polar bears, dealing first with the 49 shot ones might be both a smarter and a more viable strategy. Yet it is not the one we end up hearing about. In the debate over the climate, we often don't hear the proposals that will do the most good but only the ones that involve cutting greenhouse-gas emissions. This is fine if our goal is just to cut those gases, but presumably we want to improve human conditions and environmental quality. Sometimes greenhouse-gas cuts might be the best way to get this, but often they won't be. We must ask ourselves if it makes more sense to help 49 bears swiftly and easily or 0.06 bears slowly and expensively.

The argument in this book is simple.

1. **Global warming is real and man-made.** It will have a serious impact on humans and the environment toward the end of this century.
2. Statements about the **strong, ominous, and immediate consequences of global warming are often wildly exaggerated,** and this is unlikely to result in good policy.
3. **We need simpler, smarter, and more efficient solutions for global warming** rather than excessive if well-intentioned efforts. Large and very expensive CO_2 cuts made now will have only a rather small and insignificant impact far into the future.
4. **Many other issues are much more important than global warming.** We need to get our perspective back. There are many more pressing problems in the world, such as hunger, poverty, and disease. By addressing them, we can help more people, at lower cost, with a much higher chance of success than by pursuing drastic imate policies at a cost of trillions of dollars.

These four points will rile a lot of people. We have become so accustomed to the standard story: climate change is not only real but will lead to unimaginable catastrophes, while doing something about it is not only cheap but morally right. We perhaps understandably expect that anyone questioning this line of reasoning must have evil intentions. Yet I think—with the best of intentions—it is necessary that we at least allow ourselves to examine our logic before we embark on the biggest public investment in history.

We need to remind ourselves that our ultimate goal is not to reduce greenhouse gases or global warming per se but to improve the quality of life and the environment. We all want to leave the planet in decent shape for our kids. Radically reducing greenhouse-gas emissions is not necessarily the best way to achieve that. As we go through the data, we will see that it actually is one of the least helpful ways of serving humanity or the environment.

I hope that this book can help us to better understand global warming, be smarter about solutions to it, and also regain our perspective on the most effective ways to make the world a better place, a desire we all share.

It's Getting Hotter:
The Short Story

While global warming has many effects, like rising sea levels, melting glaciers, and maybe more ferocious hurricanes, **let us start by looking at just one factor: temperature.**

As we call it global warming, temperature is perhaps the obvious place to begin. So let's start by asking the crucial questions: What happens when temperatures increase? What can we do about this? What does it cost?

The reason we are concerned about global warming is due to the so-called greenhouse effect, the fundamental principle of which is quite simple and entirely uncontroversial. Several types of gases can reflect or trap heat, most importantly water vapor and carbon dioxide (CO_2). These greenhouse gases trap some of the heat emitted by the Earth, rather like a blanket wrapped around the globe. The basic greenhouse effect is good: if the atmosphere did not contain greenhouse gases, the average temperature on the Earth would be approximately 59°F colder, and it is unlikely that life as we know it would be able to exist.

The problem is that people have substantially increased

the quantity of CO_2 in the atmosphere, mainly from burning fossil fuels, such as coal, oil, and gas. As natural processes only slowly remove CO_2 from the atmosphere, our annual emissions have increased the total atmospheric CO_2 content—the CO_2 concentration—such that today it is 36 percent higher than in preindustrial times.

Absent a major policy change, we will continue to burn more fossil fuels over the coming century. This is especially true for the rapidly industrializing developing world, such as China and India. Whereas the developing world now is responsible for about 40 percent of the annual global carbon emissions, by the end of the century that figure will more likely be 75 percent. More CO_2 will hold on to more heat and raise temperatures. This is the man-made greenhouse effect.

Let's look at what will happen when we turn up the heat.

When talking about future climate, we obviously cannot observe it directly. Instead, researchers must make predictions about crucial factors, such as how much oil and coal each nation will use over the coming century, then put the resultant CO_2 emissions into hugely complex models that tell us how these new levels of greenhouse gases will affect the temperature, sea levels, and so forth.

Our best information comes from the United Nations' Intergovernmental Panel on Climate Change, or IPCC. Every six years or so, it gathers the best information we have on climate models and climate effects. In its "standard" future scenario, the IPCC predicts that the global temperature in 2100 will have risen on average 4.7°F from the current range. But nobody actually lives at the global average. To begin with, global warming will make land warm up faster than it will water. (It's much easier to warm up a couple of yards of earth than to warm up two deep miles of ocean.) Moreover, global warming increases *cold* temperatures much more than warm temperatures, thus it increases night and winter temperatures much more than day and summer temperatures. The reality of climate change isn't necessarily an unusually fierce summer heat wave. More likely, we may just notice people wearing fewer layers of clothes on a winter's evening. Likewise, global warming increases temperatures in temperate and Arctic regions much more than in tropical areas.

We have already experienced this warming pattern in the twentieth century. Globally, winter temperatures have increased much more than summer temperatures, and night temperatures have increased much more than day temperatures. Moreover, winter temperatures have been increasing the most in colder locations—more than three-quarters of the winter warming in the Northern Hemisphere has been confined to the very cold high-pressure systems of Siberia and northwestern North America.

Not surprisingly, this has meant that the United States, northern and central Europe, China, Australia, and New Zealand have experienced fewer frost days. However, with most warming heating up nights and winters, only Australia and New Zealand have had their maximum temperatures go up. For the United States, the maximum temperatures show no trend, and for China the maximum has even declined. In the Central England Temperature series, which is the longest record of temperatures in the world, going back to 1659, there has been a clear reduction in the number of cold days but no increase in the number of hot days.

But what will happen over the coming century, with temperatures rising 4.7°F? The standard story is that our world will become a very unpleasant one. Whenever there is a heat wave, journalists write that it might be a glimpse of much worse things to come. As one environmentalist has said, "If you don't like the current heat wave event, you're going to like it even less in the future." Famously, the chief scientific advisor to the British government, Sir David King, even envisions that an ice-free "Antarctica is likely to be the world's only habitable continent by the end of this century if global warming remains unchecked."

Nearly all discussions of the future impacts of global warming use the 2003 heat wave in Europe as their prime example. In Al Gore's words: "We have already begun to see the kind of heat waves that scientists say will become much more common if global warming is not addressed. In the summer of 2003 Europe was hit by a massive heat wave that killed 35,000 people."

Yet while we will see more and hotter heat waves, talking only about heat waves means we leave out something even more important.

Heat Deaths—Way of the Future?

The IPCC finds that the trends we have seen over the twentieth century will continue, with temperatures increasing more over land, more in the winter, and especially in the high northern latitudes: Siberia, Canada, and the Arctic. In the wintertime, temperatures might increase 9°F in Siberia compared to perhaps 5°F in Africa. There will be an increase in heat waves and a decrease in cold spells. We will see a marked decrease in frost days almost everywhere in the middle and high latitudes, and this will lead to a comparable increase in the growing-season length.

Models show that heat events we now see every twenty years will become much more frequent. By the end of the century, we will have such events happening every three years. This confirms the prospect that we could be seeing many more heat deaths—a tragedy that will indeed be caused by global warming.

But cold spells will decrease just as much as heat waves increase. In areas where there is one cold spell every three years, by the end of the century such spells will happen only once every twenty years. This means fewer deaths from cold, something we rarely hear about.

In the U.S. 2005 Climate Change and Human Health Impacts report, heat is mentioned fifty-four times and cold just once. It might seem callous to weigh lives saved versus those lost, but if our goal is to improve the lot of humanity, then it's important to know just how many more heat deaths we can expect compared to how many fewer cold deaths.

For almost every location in the world, there is an "optimal" temperature at which deaths are the lowest. On either side of this temperature—both when it gets colder and warmer—death rates increase. However, what the optimal temperature is is a different issue. If you live in Helsinki, your optimal temperature is about 59°F, whereas in Athens you do best at 75°F. The important point to notice is that the best temperature is typically very similar to the average summer temperature. Thus, the actual temperature will only rarely go above the optimal temperature, but very often it will be below. In Helsinki, the optimal temperature is typically exceeded only 18 days per year, whereas it is below that temperature a full 312 days. Research shows that although 298 extra people die each year from it being too hot in Helsinki, some 1,655 people die from it being too cold.

It may not be so surprising that cold kills in Finland, but the same holds true in Athens. Even though absolute temperatures of course are much higher in Athens than in Helsinki, temperatures still run higher than the optimum one only 63 days per year, whereas 251 days are below it. Again, the death toll from excess heat in Athens is 1,376 people each year, whereas the death toll from excess cold is 7,852.

This trail of statistics leads us to two conclusions. First, we are very adaptable creatures. We live well both at 59°F and 75°F. We can adapt to both cold and heat. Further adaptation on account of global warming will not be unproblematic, because we have already invested heavily in housing and infrastructure such as heating and air-conditioning to handle our current climate. But that is why the second point is so important. It seems reasonable to conclude from the data that, within reasonable limits, global warming might actually result in lower death rates.

Death in Europe

The heat wave in Europe in early August 2003 was exceptional in many ways. It was a catastrophe of heartbreaking proportions. With more than 3,500 dead in Paris alone, France suffered nearly 15,000 fatalities from the heat wave. Another 7,000 died in Germany, 8,000 in Spain and Italy, and 2,000 in the United Kingdom: the total death toll ran to more than 35,000. Understandably, this event has become a psychologically powerful metaphor for the frightening vision of a warmer future and our immediate need to prevent it.

The green group Earth Policy Institute, which first totaled the deaths, tells us that as "awareness of the scale of this tragedy spreads, it is likely to generate pressure to reduce carbon emissions. For many of the millions who suffered through these record heat waves and the relatives of the tens of thousands who died, cutting carbon emissions is becoming a pressing personal issue."

Such reports fueled the public perception that the heat wave became a sure indicator of global warming. But group wisdom can occasionally be wrong. A recent academic paper has checked this theory and concluded that although the circumstances were unusual, equal or more unusual warm anomalies have occurred regularly since 1979.

Moreover, while thirty-five thousand dead is a terrifyingly large number, all deaths should in principle be treated with equal concern. Yet this is not happening. When two thousand people died from heat in the United Kingdom, it produced a public outcry that is still heard. However, the BBC recently ran a very quiet story telling us that deaths caused by cold weather in England and Wales for the past years have hovered around twenty-five thousand each winter, casually adding that the winters of 1998–2000 saw about forty-seven thousand cold deaths each year. The story then goes on to discuss how the government should make more winter fuel available and how the majority of deaths are caused by strokes and heart attacks.

It is remarkable that a single heat-death episode of thirty-five thousand from many countries can get everyone up in arms whereas cold deaths of twenty-five thousand to fifty thousand a year in just a single country pass almost unnoticed. Of course, we want to help avoid another two thousand dying from heat in the United Kingdom. But presumably we also want to avoid many more dying from cold.

In Europe as a whole, about two hundred thousand people die from excess heat each year. However, about 1.5 million Europeans die annually from excess cold. That is more than seven times the total number of heat deaths. Just in the past decade, Europe has lost about fifteen million people to the cold, more than four hundred times the iconic heat deaths from 2003. That we so easily neglect these deaths and so easily embrace those caused by global warming tells us of a breakdown in our sense of proportion.

How will heat and cold deaths change over the coming century? Let us for the moment assume—very unrealistically—that we will not adapt at all to the future heat. Still, the biggest cold and heat study from Europe concludes that for an increase of 3.6°F, "our data suggest that any increases in mortality due to increased temperatures would be outweighed by much larger short-term declines in cold-related mortalities." For Britain, it is estimated that a 3.6°F increase will mean two thousand more heat deaths but twenty thousand fewer cold deaths. Indeed, a paper trying to incorporate all studies on this issue and apply them to a broad variety of settings both developed and developing around the world found that "global warming may cause a decrease in mortality rates, especially of cardiovascular diseases."

But of course, it seems very unrealistic and conservative to assume that we will not adapt throughout the twenty-first century. Several recent studies have looked at adaptation in up to twenty-eight of the biggest cities in the United States. Take Philadelphia. The optimal temperature seems to be about 80°F. In the 1960s, when it got hotter (about 100°F), the death rate increased sharply. Likewise, when temperatures dropped below freezing, it increased sharply—just like in Athens.

Yet something great happened in the decades following. Death rates dropped in general because of better health care. But crucially, temperatures of 100°F today cause almost no excess deaths. However, more people still die because of colder weather. One of the main reasons for the lowered heat susceptibility is likely increased access to air-conditioning. Studies seem to indicate that over time and with sufficient resources we actually learn to adapt to higher temperatures. Consequently, we will experience fewer heat deaths even when temperatures rise.

Moreover, we have already experienced temperature rises on the scale of what we're expecting over the coming century—in many of the major cities around the world. By the end of the century, the overwhelming majority of the world's population will likely live in cities. Cities have already experienced large temperature increases and thus provide ways for us to peek into the future and get a sense of how bad a 4.7°F increase will be.

Bricks, concrete, and asphalt, which dominate cities, absorb much more heat than vegetation does in the countryside. The effect creates what are called urban heat islands. The British meteorologist Luke Howard originally discovered the effect in the early 1800s in London, but as cities grow and replace ever more vegetation with high-rises and tarmac, we have seen the effects documented in cities around the world—from Tel Aviv, Baltimore, and Phoenix to Guadalajara in Mexico, Barrow in Alaska, Shanghai, Seoul, Milan, Vienna, and Stockholm. In downtown Los Angeles, maximum temperatures over the past century have increased some 4.5°F and minimum temperatures some 7°F. New York has a similar nighttime urban heat island of 7°F.

Recently, we have been able to use satellite measurements of the temperatures of the entire surface of a city. In looking at Houston, a fast-growing city—from 1990 to 2000, it grew by three hundred thousand residents, a full one-fifth increase—researchers found an amazing result. Over a short twelve years, the nighttime surface temperature increased about 1.4°F. Over a one-hundred-year period, that would translate into an increase of more than 12°F.

And indeed, these are the kinds of temperature differences that are being found for huge cities around the world. Asian cities are today the most rapidly growing regions of the world. Tokyo, with its twenty million inhabitants, has seen some of the most dramatic consequences of the urban heat island. While the daytime temperature of the area surrounding Tokyo in August was 83°F, the downtown was measured at above 104°F. And this high temperature is not just affecting a small inner core of the city—the high-temperature area covers some three thousand square miles, or the equivalent of 140 times the area of Manhattan. These worldwide urban temperature increases tell us at least two things.

First, many of these increases over the past half or full century are on the same order or bigger than the 4.7°F that we expect to see over the coming full century. It is likely that for many cities the urban-heat-island increases of the twentieth century are on a larger scale than those to come from global warming in the twenty-first century. Yet the increases have not brought these cities tumbling down.

This does *not* mean that such increases might not have been bad for many cities. Although deaths have in general been declining (as in Philadelphia), they might have declined even faster without it. But it means that the doomsday predictions are not objective when they fail to account for the adaptation that will possibly strongly miti-gate the temperature effects. Since our forebears were able to do so, it seems reasonable to assume that, being much richer and having vastly more technical prowess, we will be able to repeat their feat.

This also does *not* deny that with global warming the impact on cities will be considerably worse because they will be hit by a double whammy—temperature increases both from CO_2 and from still increasing urban heat islands. But unlike previous generations who did very little or noth-ing about urban heat islands, we are in a good position to alleviate many of their effects. Presumably our goal is to prevent part of the problems of increasing temperatures over the coming century.

Studies show that very simple solutions can make great differences. One of the two main reasons cities are hotter is that they are drier. Cities lack moist green spaces and have large impermeable surfaces with drainage that quickly leads any water away. Thus, the sun's energy goes into heating the atmosphere instead of into the cooling evaporation of water. If we plant trees and provide vegetation and water features in urban environments, this will—apart from making more beautiful cities—dramatically cool the surroundings. For London, that could decrease temperatures by more than 14°F.

The other main reason that cities are hotter is that they have a lot of black asphalt and heat-absorbing dark structures. Although it may seem almost comically straightforward, one of the main solutions is very simple: paint the tarmac and buildings white. Increase the general reflectivity and natural shading from buildings, and we can avoid a great deal of the heat buildup. In London, this could lower heat by 18°F.

Real-world political suggestions focus on "cool communities," reroofing and repaving in lighter colors as well as planting trees. Such a program for Los Angeles—involving planting eleven million trees, reroofing most of the five million homes, and painting one-quarter of the roads—would have a onetime cost of about $1 billion. However, it would have *annual* added benefits of lowering air-conditioning costs by about $170 million and providing $360 million in smog-reduction benefits—plus the added benefits of a greener city. And perhaps most impressive, it would lower city temperatures by more than 5°F—or about the temperature increase envisioned for the rest of this century.

The Kyoto Protocol: Buying Seven Days

But even though temperature rises may not be as devastating as people think, and even if there are other, cheap ways to deal with much of the temperature increase, is it still not also obvious that we want to cut CO_2 emissions?

Well, maybe. It really depends on how much good we can accomplish and at what cost. Let's look at this more closely.

At the moment, the only real political initiative calling for carbon cuts is the so-called Kyoto Protocol, signed in 1997. It has been championed by many environmentalists, including Al Gore, who as vice president directed the U.S. negotiations. Here it was decided that the industrial nations should reduce their overall CO_2 emissions in the period from 2008 to 2012 by about 20 percent below what they would otherwise have been.

Yet Kyoto matters little for the climate. Even if all countries had ratified it (the United States and Australia did not), and all countries lived up to their commitments (which many will have a hard time doing) and stuck to them throughout the twenty-first century (which would get ever harder), the change would have been minuscule. The temperature by 2050 would be an immeasurable 0.1°F lower and even by 2100 only 0.3°F lower. This means that the expected temperature increase of 4.7°F would be postponed just *five* years, from 2100 to 2105.

Given the centrality of Kyoto in the public discourse, it surprises most people to learn how little its adoption would actually change the future. It is worth emphasizing that these data are scientifically entirely uncontroversial. The five-year postponement over one hundred years is derived from work by one of IPCC's foremost modelers. This is why *The Washington Post* calls Kyoto a "mostly symbolic treaty." Even its staunchest backers admit Kyoto is only a small first step, and we are routinely told that we need to get much more ambitious.

Since 1997, the United States and Australia have dropped out. Countries including Canada, Italy, Portugal, and Spain are off track to meet their commitments. At the same time, Russia and other low-performing transition economies are allowed to emit much more than they do, which means that they can sell their excess emissions rights. In total, this means that Kyoto really no longer restricts CO_2. Realistically, the effective outcome will be a tiny 1 percent reduction of industrial-country emissions from what they would otherwise have been.

If no other treaty replaces Kyoto after 2012, its total effect will have been to postpone the rise in global temperature a bit less than seven days in 2100.

Why is the effect of cutting emissions so little? The answer is that the emissions from the developed world matter less and less as China, India, and other developing countries dramatically grow their economies. Yet neither China nor India seems likely to accept real limits anytime soon, basically because they have other and bigger priorities, such as food and improved living conditions. Lu Xuedu, deputy director of China's Office of Global Environmental Affairs, pointed out: "You cannot tell people who are struggling to earn enough to eat that they need to reduce their emissions." We will get back to that point.

But still, even if we reduce temperatures only a little, is that not better than doing nothing? Perhaps, but before deciding let us look at the cost.

Cost of Cutting Carbon

In estimating the cost of Kyoto, the largest assembly of the top macroeconomic models shows an average cost of about $180 billion annually from 2008. While this cost would definitely not bankrupt the industrialized world, it is still a significant amount—about 0.5 percent of global GDP. Of course, with the United States out, the cost is lower, and the effect is dramatically lower. Depending on Russia and the future of negotiations, models estimate costs as low as $5 billion to $10 billion, or nearly as high as $180 billion.

Often, people ask: Why does it cost to cut CO_2? Well, there are many different ways that cutting carbon can be costly.

Oil by itself is the most important and most valuable commodity of international trade. At more than $1.5 trillion per year, it alone accounts for more than 3 percent of global GDP. With gas and coal, fossil fuels make up more than $2.5 trillion, or 5.5 percent of global GDP. Consequently, there is big money to be made by saving energy, and incremental smart ideas in a myriad of companies and organizations every year shave off about a percent of our energy use. Actually, at least as far back as we have statistics, we have learned to produce ever more goods with the same amount of energy. Whereas the United States in 1800 produced only one present-day dollar's worth of output for a unit of energy, it today produces almost five dollars' worth.

Although a large part of the energy-efficiency increase takes place in industry and organizations, we also experience the effect as consumers. The average car driven in America has improved its mileage by 67 percent since 1973.

Likewise, home heating in Europe and the United States has improved 24 to 43 percent. Many appliances have become much more efficient—the average dishwasher, washing machine, and air conditioner have cut about 50 percent of their energy use over the past decades.

This by itself should lead us to believe that we would use less and less energy. But while the car's engine gets more efficient, we get a car with air-conditioning. While our washing machine uses less energy, we also buy a dishwasher. We heat each room more efficiently but have ever more space. While we produce each dollar's worth of goods ever more efficiently, we produce even more worth. Ingenuity still works, and people constantly find ways to cut energy use—if they didn't, our energy consumption would increase 75 percent more over the next fifty years—but our total energy consumption increases, and so do our carbon emissions.

Thus, if we want to cut emissions beyond what comes from natural ingenuity, we need to motivate people to emit less. Here, economists advocate using a tax on carbon. While nobody likes taxes (maybe apart from the secretary of the Treasury), they send a powerful signal to avoid or at least lessen carbon emissions. If you have to pay more, you will probably cut back on some of your energy consumption—maybe you will switch off the air conditioner in the car, or possibly you will even bike.

More important is what happens to industry, power plants, and heating facilities, which cause the majority of emissions. Coal is cheap but emits lots of carbon, whereas gas is more expensive but emits less, and renewables, such as biomass and solar energy, are even more expensive but

DOESN'T CUTTING EMISSIONS SAVE MONEY?

It is often claimed that we can cut emissions and actually make money. For instance, in your own home, you could lower the heat and put on a sweater or unplug some of your gizmos when they are not being used. In a recent global-warming awareness week in the United Kingdom, Tony Blair pledged to turn down his thermostat, while Sir David Attenborough promised to unplug his cell phone charger. Isn't this costless or even advantageous? After all, cutting your heating or electricity bill means money in your pocket.

Economists are typically wary of such claims. Why, they ask, would you not have done so if it really already was in your own interest to do so? Why would Sir David have to wait till the awareness week to pledge to unplug his charger if it had been an advantage for him to do so all along? This is reminiscent of Japan's Ministry of the Environment trying to catch up to the Kyoto greenhouse-gas targets by turning off the heat in February 2006:

> The ministry's "Warm Biz" campaign urges Japan's bureau-cracy and businesses to bundle up with sweaters and scarves to cut down on energy use. "It's actually not that cold. We're all keeping warm from the heat of our computers," ministry spokesman Masanori Shishido said, but admitted he has taken to wearing thermal underwear.

Likewise, we are often told about a range of energy-efficiency home improvements that we rarely do but that would actually make us money. Again, the economists wonder why we didn't already do them if they are so beneficial. For instance, engineers and manu-facturers claim that insulating your house would reduce the energy bill by 22 to 53 percent. An academic study found that the adver-tised savings were vastly overstated, thus explaining many home owners' reluctance. In the United Kingdom, a well-known environ-mental architect campaigns to get people to add a near-silent microwindmill to the gable ends of their houses. At a cost of one thousand pounds and a promise of saving up to 50 percent of a household's electricity demand, it seems like a good deal. How-ever, an independent analysis found that such a windmill would typically produce only about 5 percent of a household's electricity—one-tenth the promised amount.

emit no CO_2. With a carbon tax, businesses will tend toward cleaner but more expensive energy sources or toward processes that are more complicated but use less fuel.

This explains why cutting carbon will have real cost. It is not the tax in itself—after all, the money from taxes will be used for the public good, perhaps even for allowing other taxes to be cut. But the fact that businesses will have to use more expensive fuels or find more expensive work-arounds means that the same products and services will now be more expensive—a cost that will ultimately be borne by us, as consumers. Now, there is nothing inherently wrong about this. Since CO_2 actually does harm, you could argue that the prices before the tax didn't reflect this harm. Or to put it differently, the extra cost we pay should be compared to the environmental benefit that we receive in less global warming. This perspective is worth contemplating.

Costs and Benefits—the Value of a Ton of CO_2

Scientists, lobbyists, and politicians will tell you that we should try to solve all the world's problems—not just climate change but all the world's woes. Nevertheless, we don't. Likewise, it is often pointed out that "we have the technology" to fix much of global warming. This is true. But we also have the technology to go to the moon, yet we don't go very often. It is too expensive.

So if we don't have (or use) the resources to fix all our problems, we should think about priorities. It is obvious that we should do something about CO_2 emissions and equally obvious that we can't cut all of them—because doing so would essentially bring our society to a halt and end civilization as we know it.

One way to think about this question is to ask about one of the next tons of CO_2 we are about to put into the atmosphere. How much harm will that ton do? And how cheaply can we avoid it?

Over the next year, two soccer moms will drive their kids back and forth to school and emit a ton of CO_2. One hundred twenty-five people are going to leave their cell-phone chargers plugged in 24-7 (though presumably not David Attenborough) and over the year cause the utilities to emit an extra ton of CO_2. Three people will have hot showers for four minutes each day and add an extra ton of CO_2 to the atmosphere. In addition, many industrial and commercial processes that most of us don't even know exist will add their tons as well.

The question then is, Which ton should we cut first? That is a hard one. Maybe the soccer moms should walk? Maybe we should get the 125 guys to unplug their chargers, but how do we organize that? This is the magical part about a carbon tax—we don't actually have to find out who should cut their emissions first. Instead, the people who have the least to lose will step up first.

If we place a tax of one dollar on a ton of CO_2, gasoline will go up about one cent a gallon. Each of the soccer moms will have to pay fifty cents more each year to drive their kids to school. The electricity bill will go up, and the 125 guys will collectively pay one dollar for their cell chargers, or a little less than a cent extra each. And to shower, you will have to pay thirty-three cents extra. Now, we don't know which of these people will change their behavior slightly—

or even if anyone will. But we know that there are some people who will. Chances are they will come from industry, because a cent per gallon when you use lots gets your attention sooner.

A one-dollar tax on CO_2 will actually lead to an overall drop in emissions of a bit more than 2 percent. Remember this is over and above the 1 percent efficiency improvement that every year comes from smarter ways to use energy (the "free" reduction). So we can cut emissions. We actually have a very simple and efficient knob to turn emissions down. The temptation is to say: Why don't we turn the knob all the way up—with a thirty-dollar tax we can cut almost 40 percent?

But cutting emissions also has costs. As we saw above, industries will have to switch to more expensive fuels or more expensive procedures. Taking a shorter shower also has a cost, although in a more indirect sense. If your cost of a hot shower is ten cents per minute, after judging your finances and your preferences you might have settled on a daily four-minute shower. If a carbon tax increases this cost, you might end up deciding that you will only do three minutes instead of four. We will have gotten a CO_2 reduction, but you will also be less pleased—you will have forgone a minute under the shower that you previously judged to be worth ten cents.

In a global macroeconomic model, the total present-day cost for a permanent one-dollar CO_2 tax is estimated at more than $11 billion. So we might want to think twice about cranking up the knob to a thirty-dollar CO_2 tax, which will cost almost $7 trillion.

Essentially, the cost of turning the knob should somehow

be weighed against the benefits of less global warming. There are two ways of thinking about this, leading to somewhat the same result. We will look at cost-benefit later, but first we could think of CO_2 as pollution.

Pollution is a negative by-product of many activities. The activity itself is valuable to us, but the pollution causes problems for others. The difficulty is that we don't take these problems into account when making decisions about our activity—after all, they are not *our* problems. But if we had to pay for the damage, then those problems suddenly *would* become our problems. This is the idea behind the "polluter pays" principle. If we have to pay the price of the problem we cause, it will not necessarily mean that we will stop everything we do—after all, the activity was also valuable—but we will be forced to weigh it against the damage it will cause, and we will only make things that will do more good than harm.

This means that we must figure out the price of CO_2. How much damage does the next ton of CO_2 that you send into the atmosphere do? This is a daunting question with no definitive answer. But over the past ten years, some of the world's top natural scientists and economists have come up with a broad range of assessments that yield a fair amount of insight.

In the biggest review article of all the literature's 103 estimates, the climate economist Richard Tol makes two important points. First, the really scary high estimates typically have not been subjected to peer review and published. In his words: "Studies with better methods yield lower estimates with smaller uncertainties." Second, he finds that with reasonable assumptions the cost is very unlikely to be higher than fourteen dollars per ton of CO_2 and likely to be much smaller. When I specifically asked him for his best guess, he wasn't too enthusiastic about shedding his cautiousness—true researchers invariably are this way—but gave a best estimate of two dollars per ton.

This means that the damage we will cause by putting out one more ton of CO_2 is likely two dollars and very unlikely to be higher than fourteen dollars. Consequently, we would do best to put a two-dollar tax on CO_2 to make sure that we properly take into account the damage we cause from using fossil fuels. If we don't have a carbon tax, we end up thinking that we can pollute with CO_2 for free, even though each ton does two dollars' worth of damage. But likewise, we shouldn't overtax CO_2. If we tax it at $85, as proposed in one radical report, while the real damage is two dollars, we lose up to $83 of social benefits. This is no trivial loss—the myriad lost opportunities add up to a total onetime economic cost of more than $38 trillion. This is more than three times the annual U.S. GDP.

So setting the tax right matters. If we put it too low, we emit too much CO_2; if we put it too high, we end up much poorer without doing enough good. A crucial question, then, is the cost of cutting CO_2 under Kyoto. For the United Kingdom, the marginal cost is estimated at twenty-three dollars per ton of CO_2—between two and eleven times higher than the likely cost of climate change.

The Kyoto Protocol turns out to be too expensive compared to the good it does. We end up spending fairly large amounts of resources (up to twenty-three dollars per ton of

CO_2 averted) but doing only little good. Could we perhaps have done more good for the world with those twenty-three dollars elsewhere? The answer is yes.

Costs and Benefits of Climate Action

Most of us don't usually buy a ton of CO_2, and thus it is a bit hard to intuit whether two or twenty-three dollars is cheap or expensive. Moreover, if we were going to cut only a couple of tons, the cost probably wouldn't matter one way or the other. But as we are talking about cutting millions and even billions of tons, maybe it would make more sense to talk in total costs and total benefits. This would have the added strength of making this choice much more comparable to other choices. As participants in a democratic process, we routinely do make (collective) choices to spend billions of dollars on many public goods, including education, health, roads, and foreign aid.

The models that estimate the total costs and benefits of global warming have been around since the early 1990s, and according to the IPCC they all produce more or less the same substantial results. What is unique about these models is that they include both a climate system and an economic system, with costs to the economic system stemming from both climate changes and CO_2 cuts.

These integrated models try to incorporate the costs of all the different impacts from climate change, including those on agriculture, forestry, fisheries, energy, water supplies, infrastructure, hurricanes, drought damage, coastal protection, land loss (caused by a rise in sea level—as in Holland, for example), wetlands, human and animal survival, pollution, and migration. Costs are expressed as the sum of two quantities: the costs of adaptation (building dams, changing to other crops, and so on) and the costs we must incur from the remaining nonadapted consequences (not all land is saved by building dams; production may fall despite the introduction of new crops; and so on).

So we have a model that can take any CO_2 policy and show us *both* the economic cost of CO_2 cuts *and* the benefits (the avoided damage) from lower temperatures on agriculture, wetlands, human life, and more. This means that we can see both the costs and benefits of the Kyoto Protocol and of more stringent regulations. We can then ask, What would be the best strategy for confronting global warming?

For the full Kyoto Protocol with the United States participating, the total cost over the coming century turns out to be more than $5 trillion. There is an environmental benefit, from the slightly lower temperature toward the end of the century: about 0.3°F. The total benefit for the world comes to almost $2 trillion. Yet in total, this shows that the Kyoto Protocol is a bad deal: for every dollar spent, it does the world only about thirty-four cents' worth of good.

Perhaps tellingly, it turns out that the United States would have borne the highest cost, at almost $6 trillion, with almost trivial benefits, which by itself might explain why the United States was the least engaged party. The same pattern repeats itself for Canada and Australia, with large costs and small benefits.

Conversely, Europe has the best deal of the rich world, paying $1.5 trillion but getting almost half back in benefits. It is still not a good deal but certainly goes a long way to explain why Europe has been the most prominent backer of Kyoto.

Russia and the other transition economies would have benefited greatly from Kyoto, because they could have sold their emissions permits at a high price, to the tune of almost $3 trillion. Of course, when the time came for Western countries to pay up, it seems politically unlikely that the public in either Europe or the United States would have accepted annual transfers of more than $50 billion for what is essentially hot air.

Finally, with a net benefit of $1.4 trillion, the rest of the world is somewhat better off, with a little less than half accruing to the lowest-income countries. This benefit, however, has to be seen in the context of the rich world forking out almost $9 trillion. For every dollar spent, the rich countries do about sixteen cents of good in the developing world.

With the United States out of Kyoto, the models show that there is not much left. Essentially, Europe, Japan, and New Zealand pay $1.5 trillion, most of which will buy hot air, and a little that will buy an extremely tiny temperature change of 0.07°F by 2100.

It is perhaps worth noticing that all of the costs and benefits above assume that policies are implemented globally and efficiently—that the smartest policies are used and coordinated globally, given the required reductions. Thus, should the rich countries not use Russia's extra quotas, the costs could escalate to almost double with virtually no added benefit. And it requires that policy makers cut CO_2 where it is absolutely cheapest—if not, the total cost has essentially no upper limit. In the United Kingdom, the

Kyoto CO_2 tax ought to be about five dollars per ton after the United States has left. However, the United Kingdom requires electricity generators to supply 10 percent of its electricity from certified renewable sources, under the so-called Renewables Obligation. Currently, that cost is estimated at $169, or more than thirty times too expensive. And remember that the Kyoto tax is probably too high compared with the CO_2-damage cost of about two dollars per ton. Thus, for every dollar spent on the Renewables Obligation, the United Kingdom gets three cents back on its Kyoto promise and does about one penny's worth of good.

When confronted with the point that doing Kyoto is an extraordinarily expensive way of doing very little good far into the future, many react by saying we should do much more. But doing more of a bad deal is rarely smart, and the models show us just that. Cutting more carbon is likely to cost ever more while doing ever less extra good. And it is still worth remembering that the cost assumes politicians will choose the very smartest tools available.

That cutting more gets ever more costly also makes good intuitive sense. If we cut a little, it is easy—surely anything you do could be improved relatively cheaply to emit a little less CO_2. The metaphor is that we pick the low-hanging fruit first, which is easily accessible. However, as we try to cut more and more, it will get ever more expensive—we have to reach high into the tree for the remaining, scarce fruit. Also, the first cuts we make mean we cut into the highest, most scary part of the temperature increases. As we cut ever deeper, we will start to cut into more usual temperatures. So while the cost increases with deeper cuts, the benefit decreases.

This is evident when we move from Kyoto to some of the more ambitious policies contemplated. Stabilizing the temperature increase to 4.5°F does more good—it reduces temperatures 0.86°F—but at a rather high cost of $15.8 trillion. It is instructive to compare this to the total cost of global warming. Models show that if global warming wasn't happening, the world would end up about $14.5 trillion richer. Thus, the cost of global warming can be estimated at $14.5 trillion. Stabilizing the temperature increase at 4.5°F actually means we end up paying more for a partial solution than the cost of the entire problem. That is a bad deal.

The most ambitious plan is to stabilize the temperature rise at 2.7°F. This is essentially the stated preference of the EU. The EU has reconfirmed this decision over the last many years. One central model shows it is possible to achieve such very low temperature increases, but only at the formidable cost of $84 trillion. For every dollar spent, it will do thirteen cents' worth of good.

Actually, the models also show a reduction that does more good than it costs. This initiative sets a global carbon tax that balances with the future environmental benefits from CO_2 cuts. It starts off with a carbon tax around two dollars per ton, rising to about twenty-seven dollars at the end of the century, reflecting how damages rise with more CO_2 in the atmosphere. The total climate impact is rather small—it only reduces the temperature increase by 0.2°F by the end of the century. Uniquely, it costs about $600 billion but creates twice that in benefits, meaning for each dollar it does two dollars of social good.

This result is surprising and runs counter to most of the climate-change proposals. We try to cut CO_2 through Kyoto, but in reality it is a poor use of resources. Many, including the EU, think we need to go much further, but the economic models show us that this is likely to be an even poorer use of resources. In general, this shows that we need to be very careful in our willingness to act on global warming. Going much beyond the small optimal initiative is economically unjustified. And this conclusion does not come from the output just of a single model.

All major peer-reviewed economic models agree that little emissions reduction is justified. A central conclusion from a meeting of all economic modelers was: "Current assessments determine that the 'optimal' policy calls for a relatively modest level of control of CO_2." In a review from 2006, the previous research was summarized: "These studies recommend that greenhouse-gas emissions be reduced below business-as-usual forecasts, but the reductions suggested have been modest."

This is such a robust result because the economic cost comes up front, whereas the benefit comes centuries down the line. If we try to stabilize emissions, it turns out that for the first 170 years the costs are greater than the benefits. Even when the benefits catch up in the late twenty-second century, there is still a payback time before the total benefits outweigh the total costs, around 2250. Thus, as one academic paper points out, "the costs associated with an emissions stabilization program are relatively large for current generations and continue to increase over the next 100 years. The first generation to actually benefit from the stabilization program is born early during the 24th century." If our desire is to help the many generations that come before then, along with the world's poor, cutting emissions is not the best way. Perhaps more surprisingly, cutting emissions is also not the best way to help people in the twenty-fourth century, since we could have focused on solving many other immediate problems that would leave the far future much better off.

We clearly need smarter ways to deal with climate change.

Living in a Hotter World

But of course, all this economist talk doesn't help if the whole world collapses. If many or most of the planet's inhabitants die from increased heat deaths, what does it matter that lower taxes are cheaper?

As it turns out, such understanding is widespread, but the threat is vastly exaggerated. And not just for the United Kingdom, as we saw above. The first complete survey for the world was published in 2006, and what it shows us very clearly is that climate change will not cause massive disruptions or huge death tolls. Actually, the direct impact of climate change in 2050 will mean *fewer* dead, and not by a small amount. In total, about 1.4 million people will be saved each year, due to more than 1.7 million fewer deaths from cardiovascular diseases and 365,000 more deaths from respiratory disorders. This holds true for the United States and Europe (each with about 175,000 saved), as for the rest of the industrialized world. But even China and India will see more than 720,000 saved each year, with deaths avoided outweighing extra deaths nine to one. The only region where deaths will outweigh lives saved is in the rest of the developing world, especially Africa. There almost

200,000 deaths will be avoided, but more than 250,000 will die.

The reaction of both my editor and several of my friends on reading this chapter was telling—and perhaps also similar to your reaction: Yes, but what happens after 2050—will heat deaths not eventually begin outweighing avoided cold deaths? It is a good question. But the survey shows that the result doesn't hold true just in 2050: in the central estimates of the model, *lives saved will continue to outweigh extra deaths when counting both cardiovascular and respiratory diseases at least till 2200.* So the simple answer to the question is no, heat deaths will not outweigh avoided cold deaths, not in 2050, 2100, or even 2200.

But it is just as important that you (well, certainly my friends and my editor) instinctively felt the need to question the claim that global warming will have positive impacts. We do not seem to have the same inquisitiveness when we're told about negative impacts. We will get back to this when discussing malaria.

Now we can also answer what we can and should do. Many commentators point out how global warming will hit the developing world hardest. For deaths from heat and cold, this is true. But most commentators then go on to suggest that we should cut carbon emissions, possibly dramatically, to help the third world. This, however, turns out to be very questionable advice.

If we got everyone on board for Kyoto, including the United States, we would see reduced temperatures in 2050 by about 0.1°F, or a postponement of temperature increases by less than three years. Thus, the developing world would experience slightly less warming by 2050. Outside of China and India, the consequence would be about 6 percent fewer deaths from respiratory disorders, saving more than fifteen thousand lives per year that would otherwise have ended. This seems to confirm that CO_2 reductions are the way to go.

But first, we have to remember that the world would pay about $180 billion annually for fifty years before this benefit would occur. Roughly that means paying almost $100 million for each life saved, which is much more than what we pay to save lives in the developed world and certainly vastly more expensive than what lives can be saved for in the developing world—estimated at less than two thousand dollars.

Second, we have to remember that cutting CO_2 and temperature means more people will die from cold deaths in the developing world—more than eleven thousand annually. So in reality, we have saved only a little fewer than four thousand people (now at the price of $300 million each).

Third, we cannot just change the thermostat where temperature impact is on the whole negative. We also end up changing it in all the other places—in the United States, Europe, Russia, China, and India. Here the effect of Kyoto on lives lost directly from temperature impacts would mean an increase in deaths of about 88,000 annually. Thus, to save four thousand people in the developing world, we end up sacrificing more than $1 trillion and eighty thousand people. Bad deal.

Global Warming Is Not the Only Issue

This does *not* mean that we should do nothing and just embrace global warming. We have looked at just one aspect: direct impacts from temperature. There are other issues, which we will discuss later, in which general and long-term impacts will be more negative.

But this aspect tells us three things, loud and clear. First, our understanding of global warming as shaped by the media and the environmental pundits is severely biased. We are being told by respected scientist James Lovelock that with the coming climate-change devastation the thirty-five thousand dead in Europe in 2003 was just the prelude to a new Stone Age: "Before this century is over billions of us will die and the few breeding pairs of people that survive will be in the Arctic where the climate remains tolerable." This is far beyond the pale of our understanding of climate change. And getting the facts wrong means we may make staggeringly bad policy judgments.

Second, when we talk about global warming, we seem obsessed with regulating just one parameter—namely, CO_2. But while turning down the CO_2 knob may be part of the solution, surely our primary concern ought to be to advance human and environmental well-being the most, where many other knobs are in play. While cutting CO_2 will save some people from dying of heat, it will simultaneously cause more people to die from cold. This highlights how reducing CO_2 means indiscriminately eliminating both negative and positive effects of global warming. We ought at least to consider adaptive strategies that would allow us to hold on to the positive effects of climate change while re-ducing or eliminating its damages.

Third, global warming is not the only issue we need to tackle. This especially holds true for the third world. Take a look at the World Health Organization's estimates of what kills us. WHO finds that climate change kills about 150,000 people in the developing world now, but as we will see in the section on malaria below, they failed to include avoided cold deaths, so this estimate is vastly overstated.

Nevertheless, it is obvious that there are many other and more pressing issues for the third world, such as almost four million people dying from malnutrition, three million from HIV/AIDS, 2.5 million from indoor and outdoor air pollution, more than two million from lack of micronutri-ents (iron, zinc, and vitamin A), and almost two million from lack of clean drinking water.

Even if global warming exacerbates some of these prob-lems (as we will discuss), it is important to point out that the total magnitude of the problems is likely to far exceed the extra problems caused by climate change. Thus, poli-cies reducing the total problem will benefit many more people than policies that only reduce the small additional part caused by global warming. Again, are there better ways to help than by cutting CO_2?

We have to ask ourselves, What do we want to do first? This is a question I have been deeply involved in answering through a process called the Copenhagen Consensus. We asked some of the smartest economists in the world, Where do extra resources do the most good first? For each prob-lem, we asked experts to put forward the very best solutions. With global warming, this might be CO_2 taxes or Kyoto, whereas for malnutrition it might be agricultural research, and for malaria it could be mosquito nets. But each expert didn't just state that the solutions do good— they showed just how much good they would do and how much they would cost.

In essence, the experts assessed the dollar value of different solutions, just as we saw them doing for climate. In that case, they estimated the benefit of Kyoto for each of the positive impacts on agriculture, forestry, fisheries, water supply, hurricane damage, and so on. They estimated the costs through losses in production. For malaria solutions, the positive effects would be the value of fewer dead, fewer sick, fewer absences from work, more robust populations with respect to other diseases, and increased production. The costs would be the dollar amount spent on buying, distributing, and using mosquito nets.

A panel of top-level economists, including four Nobel laureates, then made the first explicit global priority list ever, shown in table 1. It divided the world's opportunities into "very good," "good," and "fair" according to how much more good they would do for each dollar spent. "Bad" opportunities were those where each dollar would do less than a dollar's worth of good.

Some of the top priorities also correspond to some of the top risk factors identified by WHO. Preventing HIV/AIDS turns out to be the very best investment humanity can make: each dollar spent on condoms and information will do about forty dollars' worth of social good (the value of fewer dead, fewer sick, less social disruption, and so on). For

	CHALLENGE	OPPORTUNITY
VERY GOOD OPPORTUNITIES	1 Diseases	Control of HIV / AIDS
	2 Malnutrition	Providing micronutrients
	3 Subsidies & Trade	Trade liberalization
	4 Diseases	Control of malaria
GOOD OPPORTUNITIES	5 Malnutrition	Development of new agricultural technologies
	6 Sanitation & Water	Small-scale water technology for livelihoods
	7 Sanitation & Water	Community-managed water supply and sanitation
	8 Sanitation & Water	Research on water productivity in food production
	9 Government	Lowering the cost of starting a new business
FAIR OPPORTUNITIES	10 Migration	Lowering the barriers to migration for skilled workers
	11 Malnutrition	Improving infant and child nutrition
	12 Malnutrition	Reducing the prevalence of low birth weight
	13 Diseases	Scaled-up basic health services
BAD OPPORTUNITIES	14 Migration	Guest workers programs for the unskilled
	15 Climate	Optimal carbon tax ($25–$300)
	16 Climate	Kyoto Protocol
	17 Climate	Value-at-risk carbon tax ($100–$450)

TABLE 1 Global priority list for spending extra resources, from the 2004 Copenhagen Consensus.

$27 billion, we can save twenty-eight million lives over the coming years.

Malnutrition kills almost four million people each year. Perhaps even more dramatically, it affects more than half the world's population, by damaging eyesight, lowering IQ, reducing development, and restricting human productivity. Investing $12 billion could probably half the incidence and death rate, with each dollar doing more than thirty dollars' worth of social good.

Ending first-world agricultural subsidies and ensuring free trade would make almost everyone much better off. Models suggest that benefits of up to $2.4 trillion annually would be achievable, with half of that benefit accruing to the third world. In achieving this, it would be necessary to pay off first-world farmers, who are used to the benefits of a closed market, but the benefits of each dollar would do more than fifteen dollars' worth of social good.

Finally, malaria takes more than one million lives each year. It infects about two billion people every twelve months (many several times) and causes widespread debilitation. Yet an investment of $13 billion could cut its incidence by half, protect 90 percent of newborns, and cut deaths of children under five by 72 percent. For each dollar spent we would do at least ten dollars' worth of social good—a compellingly sound investment, especially when so many of the lives saved would come from nations that carry the weight of the world's greatest problems.

At the other end of the spectrum, the Nobel winners placed climate-change opportunities, including Kyoto, at the bottom under the heading "bad opportunities," underlining what we saw above—namely, that for each dollar spent, we would end up doing much less than a dollar's worth of good for the world.

But the Copenhagen Consensus did not just ask top economists their opinions. After all, economists (fortunately!) don't lead the world. We asked future leaders—eighty college students from all over the world, with 70 percent from developing countries, with equal gender representation, and from the arts, sciences, and social sciences—what they thought. After five days of meeting with world-class experts on each of the ills, these bright young minds came to a surprisingly similar result. They placed malnutrition and communicable diseases on top, climate change next to last.

In 2006, we ran the project again, asking a wide range of UN ambassadors to make their priority list. Besides the three biggest countries—China, India, and the United States—countries as diverse as Angola, Australia, and Azerbaijan participated, along with Canada, Chile, Egypt, Iraq, Mexico, Nigeria, Poland, Somalia, South Korea, Tanzania, Vietnam, Zimbabwe, and many others.

Politicians don't really like to make choices—they don't like to pick between two groups: they'd rather pretend they can do it all. After I briefed the ambassadors, several of them looked down the list of major challenges facing the planet and said, "I want to put a number one by each of these projects." But prioritize they did. They came out with a list similar to the economists', placing communicable diseases, clean drinking water, and malnutrition at the top, with climate change toward the bottom.

This should make us pause. None of these forums has said that climate change is fictitious or not important. What they've done is ask us to consider addressing the real and pressing needs of current generations that we can solve so easily and cheaply before we try to tackle the long-term problem of climate change, which will be massively expensive and accomplish so little.

Of course, it's tempting for us to say we should do it all. Morally, that seems compelling. But the truth is that's not realistic. The world lacks the resources and will to solve all its major challenges. Focusing on some issues puts others on the back burner.

These forums also point out that in addressing these immediate problems we do more than just fix the problems of today. Imagine dramatically reducing HIV/AIDS and malaria. Picture a world where more than half the world's population doesn't succumb to developmental deficiencies from malnutrition. Envision a time when rich world agriculture isn't heavily subsidized, actually allowing the third world to sell its products in first-world markets. This is a world I would like to live in. It would not just give communities an immediate boost but also would leave them strengthened, with economies growing faster, meaning still stronger and richer societies by 2100. It will enable these societies to deal much better with future problems—be they natural or from global warming. Investing in today will not only alleviate today's problems but also improve tomorrow.

When we look into the future, the UN expects that people in both the developed and developing countries will become richer. In the industrialized world, people will see their incomes grow sixfold, as we saw during the last century. Income in the developing countries is expected to soar twelvefold.

This is important when talking about climate change. In the UN's most likely scenario for 2100, when many of warming's problems will be felt in earnest, the average person in the developing world is expected to make about one hundred thousand dollars (in present value) each year. Even the very worst-case scenario envisions the average person making above $27,000. In this very unlikely case, the average person in the third world will be as rich as a present-day Portuguese or Greek or richer than most West Europeans in 1980. Much more likely, he or she will be richer than today's average American, Dane, or Australian. This richness will of course enable these countries to better handle outside shocks, whether they come from climate change or any of the other major challenges the future undoubtedly will deal us.

And this is the final part of the reason Nobelists, youngsters, and UN ambassadors place disease and malnutrition so high and climate so relatively low. When we try to help the developing world by cutting our carbon emissions, we are trying to help people far into the future, where they will be much richer. We are not helping a poor Bangladeshi in 2100 but much more likely a rather rich Dutchman. And in case you wonder whether global warming will mean that Bangladesh will be underwater in 2100, we will see below that a rich Bangladesh will lose only 0.000034 percent of its present dry-land area.

The question then becomes whether we wouldn't do better by helping a poor Bangladeshi today. He or she needs our help more, and we can do much more for him or her. Helping a present-day Bangladeshi become less sick, better fed, and better able to participate in the global marketplace will not just do obvious good. It will also enable him or her to better Bangladeshi society, grow the Bangladeshi economy, and leave a richer and more robust Bangladesh to future generations, who will be much better equipped to deal with global warming. To me, that's a compelling case for action.

Our Generational Mission

Yet it is clear that many people in the developed world feel differently. In one recent survey in Australia, environmental concern came in as absolutely the most important priority for the leaders of the world, before eliminating poverty or dealing with terrorism, human-rights issues, and HIV/AIDS. In another survey, the United States, China, South Korea, and Australia found improving the global environment a more important foreign-policy goal than combating world hunger. South Korea put it first on its list of the top sixteen global threats.

Why are we so singularly focused on climate change when there are many other areas where the need is also great and we could do so much more with our effort?

Al Gore gives us two reasons. First, it is a planetary emergency: "At stake is the survival of our civilization and the habitability of the Earth." Yet this turns out to be far from the truth. As we saw above, this is not what the science is telling us for the temperature rise over the coming century. If anything, the science tells us that *fewer* people will die with moderately more heat. Of course, Gore has several other arguments, which we will also address below.

Gore's second reason is probably more telling and closer

to the truth. He tells us how global warming can give *meaning* to our lives.

> The climate crisis also offers us the chance to experience what very few generations in history have had the privilege of knowing: *a generational mission;* the exhilaration of a compelling *moral purpose;* a shared and unifying *cause;* the thrill of being forced by circumstances to put aside the pettiness and conflict that so often stifle the restless human need for transcendence; *the opportunity to rise.* . . . When we rise, we will experience an epiphany as we discover that this crisis is not really about politics at all. It is a moral and spiritual challenge.

He explains how global warming can give us a moral imperative, like the one Lincoln had for fighting slavery or Roosevelt had against fascism or Johnson had for the rights of minorities.

It seems unrealistic to expect that climate change will give us such singularity of purpose. If anything, the ten-year drawn-out battles around the relatively minor restrictions of Kyoto show us that anything costing individual nations trillions of dollars will be strongly contested and lead to strife rather than serenity.

But perhaps more important, should we go for the exhilaration of a generational mission just because it makes us *feel* good? Should it not actually be because we are *doing* the best our generation can do? And this, of course, brings us right back to asking whether there are greater opportunities for us to engage first.

To be fair, Gore does point out that there are many other generational missions:

> The understanding we will gain [from tackling climate change] will give us the moral capacity to take on other related challenges that are also desperately in need of being redefined as moral imperatives with practical solutions: HIV/AIDS and other pandemics that are ravaging so many; global poverty; the ongoing redistribution of wealth globally from the poor to the wealthy; the ongoing genocide in Darfur; the ongoing famine in Niger and elsewhere; chronic civil wars; the destruction of ocean fisheries; families that don't function; communities that don't commune; the erosion of democracy in America; and the refeudalization of the public forum.

But as the list goes on, it becomes clear that it is in need of realistic prioritization. Gore essentially tells us we should fix all things from climate change to democracy. And it would be beautiful if we could do so. But so far, we haven't addressed any of these very well. Perhaps it would be wise to start thinking about which we should do first.

Gore tells us that we need to hear the voices of the future speaking to us now. We have to imagine them asking: What were you thinking? Didn't you care about our future? He is absolutely right.

Do we want future generations to say that we have spent trillions of dollars and perhaps done a little good for rich people in a hundred years? Or do we want future generations to thank us for giving billions of poor people a new beginning and a better life, which will enable them to better deal with whatever challenges the future holds?

In other words, do we just want to *feel* good, or do we actually want to *do* good?

Smarter Strategies

This does not mean we should do nothing at all about climate change. It does mean we need to be much smarter about what we do. In place of expensive and inefficient strategies like Kyoto, we should search for new solutions, which I will discuss below.

We should start thinking about how we can cheaply tackle climate change in the long run. The big problem with cutting carbon emissions Kyoto-style is that it costs a lot now and does very little, far into the future. And as long as cutting carbon has such high costs, countries such as China and India will never participate.

Rather we should focus on cutting the cost of reducing CO_2 emissions. While this may not be as romantic as a "generational mission," it is much more efficient, stands a much better chance of working, and will really help humanity.

Global Warming:
Our Many Worries

In chapter 2, we looked at what happens just when temperatures increase and saw that it was no catastrophe. But of
course, there are many other concerns about global warming, each often presented as a disaster-in-waiting, urging us
to drop everything else and focus on cutting CO_2. As it turns
out, these statements are often grossly exaggerated and
divert us from making sound policy judgments. Let's take a
realistic look at some of them.

Melting Glaciers

Over the past millennium, temperatures have gone up and
down and up again from natural causes. (In the past 150
years, temperatures have diverged even more upward due
to global warming.) Between about 900 and 1200 there was
a relatively warmer period known as the Medieval Warm
Period. The warmer climates and reduced sea ice made
possible the colonization of the otherwise inhospitable
Greenland and Vinland (Newfoundland) by the Vikings. In

Alaska, the mean temperature was 3 to 5°F warmer in the eleventh century than today, and the snow line in the Rocky Mountains was about three hundred yards higher than today.

Likewise, the middle of the last millennium saw a marked cooling, named the Little Ice Age. Evidence from a wide range of sources shows colder continents. Glaciers advanced rapidly in Greenland, Iceland, Scandinavia, and the Alps. The Arctic pack ice extended so far south that there are six records of Eskimos landing their kayaks in Scotland. Many European springs and summers were outstandingly cold and wet, and crop practices changed throughout Europe to adapt to a shortened and less reliable growing season, causing recurrent famines. The harsh winters were captured by the Flemish artist Pieter Brueghel (1525–69), who initiated a genre by completing at least seven winter landscapes in two years. Possibly the worst winter in France, in 1693, is estimated to have killed several million people— about 10 percent of the population. Likewise in China, warm-weather crops, such as oranges, were abandoned in the Kiangsi Province. In North America, the early European settlers reported exceptionally severe winters, with Lake Superior iced over till June in 1608.

It is perhaps worth noticing that while many events during the Little Ice Age are seen and reported as negative, this does not seem to be the case with most of the Medieval Warm Period.

It is clear that part of the temperature increase since then has simply been a result of coming out of the Little Ice Age. It is also clear, though, that we are now seeing a warming trend beyond that, indicating man-made global warming.

Both of these warmings have caused glaciers to recede. Many have seized on pictures of these retreating glaciers as symbols of global warming. Al Gore, for example, fills eighteen pages of his book with before-and-after pictures of glaciers.

But several facts impede this rather simple narrative. First, glaciers have been greatly advancing and receding since the last ice age. In Switzerland, there have been twelve such advances and retreats over the past ten thousand years. One of the best-studied glaciers in Norway—my namebrother Bjornbreen—was entirely absent for three thousand years over two periods about seven thousand years ago; over the last ten thousand years, it has been reborn six times. In fact, most glaciers in the Northern Hemisphere were small or absent from nine thousand to six thousand years ago. While glaciers since the last ice age have waxed and waned, they overall seem to have been growing bigger and bigger each time until reaching their absolute maximum at the end of the Little Ice Age. It is estimated that glaciers around 1750 were more widespread on Earth than at any time since the ice ages twelve thousand years ago. When Bjørnbreen peaked around 1800, it was actually twice as large as in any of its five previous incarnations.

So it is not surprising that as we're leaving the Little Ice Age we are seeing glaciers dwindling. We are comparing them with their absolute maximum over the past ten millennia. The best-documented overview of glaciers shows that they have been receding continuously since 1800. The perfect glacier icon, the snow-clad Mount Kilimanjaro, has been receding at least since 1880. When Ernest Hemingway published "The Snows of Kilimanjaro" in 1936, the mountain had already lost more than half its glacier surface area in the previous half century. This is more than it has lost in the seventy years since. Actually, the central theme from the inception of published research on Kilimanjaro in 1891 has been the drastic recession of its glaciers.

Furthermore, Kilimanjaro has not lost its ice on account of increasing temperatures, which have remained rather stable below freezing, but because of a regional shift around 1880 toward drier climates. Thus, Kilimanjaro is not a good poster child for man-made global warming. In the latest satellite study, it is concluded that "results suggest glaciers on Kilimanjaro are merely remnants of a past climate rather than sensitive indicators of 20th century climate change."

Yet we are often told that we need to reduce CO_2 emissions to address the problem of the receding glaciers. In a video with Kilimanjaro in the background, Greenpeace tells us that the mountain's entire ice field might be lost by 2015 due to climate change: "This is the price we pay if climate change is allowed to go unchecked." But of course, for Kilimanjaro we are able to do nothing, since it is losing ice due to a drier climate. Even if we granted that its demise was partially related to global warming, nothing we could do would have even the slightest impact before 2015.

When Greenpeace informs us that "Mount Kilimanjaro's fast-melting glaciers symbolise the fact that climate change may be felt first and hardest by the environment and people of Africa," it is mixing two messages. No, Kilimanjaro is not a good symbol of climate change, but yes, climate change will definitely hit harder in the developing world.

While emotionally charged pictures of the beautiful glaciers from Kilimanjaro paired with admonishing concerns over CO_2 undoubtedly are very effective with the media and opinion makers, they hardly address the real problems of the Tanzanian farmers on the slopes. I don't claim to know the concerns of Tanzania any better than Greenpeace does. But what I do know is that local surveys in that country show the biggest concerns are the lack of capital to buy seeds, fertilizers, and pesticides; pests and animal diseases; costly education; high HIV-infection rates; malaria; and low-quality health services. I believe we have to dare to ask whether we help Tanzanians best by cutting CO_2, which would make no difference to the glaciers, or through HIV policies that would be cheaper, faster, and have much greater effect.

Yet there is a real concern when it comes to glaciers that feed rivers, such as the Himalayan glaciers on the Tibetan plateau. They are the biggest ice mass outside of Antarctica and Greenland, and they are the source of rivers that reach 40 percent of the world's population. A stable glacier is not a continuous source of water into the rivers (if it were, it would quickly vanish), but it acts as a seasonal water tower, accumulating water as ice in the winter, then releasing it in the summer. In this way, melting glaciers provide as much as 70 percent of the summer flow in the Ganges and 50 to 60 percent of the flow in other major rivers.

The concern is that if glaciers entirely disappear, the overall amount of water available over a year would probably remain the same (with roughly the same amount of precipitation), but it would be distributed very differently, possibly leading to severe summer droughts. To a large extent this can be remedied by improved water storage, but of course that would mean large extra costs.

However, there are two important points to be made. First, with glacial melting, rivers actually *increase* their water contents, especially in the summer, providing *more* water to many of the poorest people in the world. Glaciers in the Himalayas have been declining significantly since the end of the Little Ice Age and have caused increasing water availability throughout the last centuries, possibly contributing to higher agricultural productivity. But with continuous melting, the glaciers will run dry toward the end of the century. Thus, global warming of glaciers means that a large part of the world can use more water for more than fifty years before they have to invest in extra water storage. These fifty-plus years can give the societies breathing space to tackle many of their more immediate concerns and grow their economies so that they will be better able to afford to build water-storage facilities.

Now, you may be cringing and saying that we should conserve the pristine glaciers. They are aesthetically magnificent, and in the best of all worlds, where there are no competing demands, that would be an important ideal. But in a world of many other issues, we have to consider that developing countries might be interested in using some of their finite natural resources, like glaciers, to grow richer rather than to provide aesthetic enjoyment for the wealthy. Certainly we did so in the developed world when we cut down much of our forests and grew rich. Moreover, it is perhaps worth contemplating the alternative. Since 1961 in the Karakoram and Hindu Kush mountains of the Upper Indus

Basin, summer temperatures have been bucking the global-warming trend, cooling almost 2°F. The consequence has been a reassuring thickening and expansion of the Karakoram glaciers, but—predictably—the summer runoff in the rivers Hunza and Shyok has decreased about 20 percent. This is important, as the two provide more than 25 percent of the inflow to the Tarbela Dam, which controls one of the world's largest integrated irrigation networks. It boils down to a stark choice: Would we rather have more water available or less?

Second, what can we actually do? We could implement Kyoto at very high costs but make virtually no impact on the Himalayan climate by 2050. Glaciers have been receding for several centuries, accumulation rates have dropped since 1840 due to a change in the Pacific trade winds, and even bold and quick CO_2 cuts will change this very little. But apart from all the other more urgent needs we should address, we could also do much more good by investing in better water infrastructure, so that India and China could better utilize the extra water now and prepare for when the rivers revert to "normal" water flow, but with more in the winter, less in the summer.

While we often hear worries about how melting glaciers will lead to less water later, we seldom hear that it is a boon now. And when it is advocated that we instantly turn the big, hard knob of CO_2 cuts, which will do little to save the glaciers at very high costs, we should be asking whether there are other, nimbler, more efficient, and less expensive knobs to turn first, where we can help the world much more effectively.

Rising Sea Levels

Another of the most doom-laden impacts from global warming is the rising sea levels. This worry is perhaps not surprising, since from time immemorial most cultures have had legends of catastrophic floods, which covered the entire Earth and left few animals and plants alive. In Western societies, the most famous version is the story of Noah saving what he could in his ark. Many commentators powerfully exploit this biblical fear of flooding, as when Bill McKibben said of our responsibility for global warming that "we are engaging in a reckless drive-by drowning of much of the rest of the planet and much of the rest of creation."

When sea levels rise, it is not due to sea ice melting, because it is already displacing its own weight—if you put ice cubes in a glass of water, the waterline will not change when they melt. Thus, contrary to common statements, the Arctic melting will not change sea levels. Instead, sea levels rise due to two factors. First, when water gets warmer, like everything else it expands. Second, runoff from land-based glaciers adds to the ocean water volume. Over the past forty years, glaciers have contributed about 60 percent and water expansion 40 percent of the rise in sea levels.

In its 2007 report, the UN estimates that sea levels will rise about a foot over the rest of the century. While this is not a trivial amount, it is also important to realize that it is certainly not outside historical experience. Since 1860, we have experienced a sea-level rise of about a foot, yet this has clearly not caused major disruptions. It is also important to realize that the new prediction is *lower* than the previous IPCC estimates and much lower than the expectations from the 1990s of more than two feet and from the 1980s, when the Environmental Protection Agency projected more than six feet.

Often, the risk of sea-level rise is strongly dramatized in the public discourse. A cover story of *U.S. News & World Report* famously predicted that "global warming could cause droughts, disease, and political upheaval" and other "nasty effects, from pestilence and famine to wars and refugee movement." We will return to these concerns later, but their primary projection for sea-level rise was this: "By midcentury, the chic Art Deco hotels that now line Miami's South Beach could stand waterlogged and abandoned."

Yet sea-level increase by 2050 will be about five inches—no more than the change we have experienced since 1940 and less than the change those Art Deco hotels have already stood through. Moreover, with sea-level changes occurring slowly throughout the century, economically rational foresight will make sure that protection will be afforded to property that is worth more than the protection costs, and settlement will be avoided where costs will outweigh benefits. The IPCC cites the total cost for U.S. national protection and property abandonment for more than a three-foot sea-level rise (more than triple what is expected in 2100) at about $5 billion to $6 billion over the century. Considering that the adequate protection costs for Miami would be just a tiny fraction of this cost spread over the century, that the property value for Miami Beach in 2006 was close to $23 billion, and that the Art Deco National Historic District is the second-largest tourist magnet in Florida after Disney World, contributing more than $11 billion annually to the economy, five inches will simply not leave Miami Beach hotels waterlogged and abandoned.

But this of course is exactly the opposite of what we often hear. In a very moving section of the film *An Inconvenient Truth,* we see how large parts of Florida, including all of Miami, will be inundated by twenty feet of water. We see equally strong clips of San Francisco Bay being flooded, the Netherlands being wiped off the map, Beijing and then Shanghai being submerged, Bangladesh being made uninhabitable for sixty million people, and the deluging of even New York City and its new World Trade Center memorial.

How is it possible that one of today's strongest voices on climate change can say something so dramatically removed from the best science? The IPCC estimates a foot, Gore tops them twenty times. Well, technically, Al Gore is not contradicting the UN, because he simply says: "If Greenland melted or broke up and slipped into the sea—or if half of Greenland and half of Antarctica melted or broke up and slipped into the sea, sea levels worldwide would increase by between 18 and 20 feet." He is simply positing a hypothetical and then in full graphic and gory detail showing us what—hypothetically—would happen to Miami, San Francisco, Amsterdam, Beijing, Shanghai, Dhaka, and then New York City.

Gore is correct in identifying Antarctica and Greenland as the most important players if he is to support his hypothetical twenty feet. The UN estimates that over the century by far the largest contribution to sea-level rise will be warmer water expanding—this alone will constitute nine of the almost twelve inches by 2100. Melting glaciers and ice caps will contribute a bit more than three inches over the century. Likewise, Greenland is expected to contribute 1.4 inches by itself. This adds up to 13.5 inches over the coming century. However, as the world warms, Antarctica will not noticeably start melting (it is still way too cold), but because global warming also generally produces more precipitation Antarctica will actually be accumulating ice, *reducing* sea levels by two inches. Thus, the total estimate of about one foot.

So where are the nineteen missing feet? All climate models are in fair agreement on the larger part of the expanding water, so it is unlikely that they are there. Likewise, they are not in the melting glaciers, because even if all glaciers and ice caps entirely disappeared, they would still contribute maximally one foot. However, if Greenland melted, that would contribute twenty-four feet. If Antarctica entirely slipped into the ocean, it would contribute a whopping 186 feet.

But will they? The available analyses tell us a very different story. Let us look at them individually, since Greenland and Antarctica are very different. Whereas the Antarctic is surrounded by ice shelves and located in a region where little or no surface melting occurs, Greenland is situated in a region where temperatures are high enough to cause widespread summer melting.

Greenland is, as the IPCC pointed out, experiencing a small overall mass loss. Some analyses have shown more rapid loss in recent years (2002–5), but by early 2007 two of the major glaciers in Greenland were again seen reverting to much lower rates of ice-mass loss. Even with the most extreme estimates of Greenland melting over a couple of years, a sea-level rise of twenty feet would take one thousand years. In a recent overview of all the major models of sea-level increase, Greenland's contribution over the coming century is at most two inches. Some even posit a tiny decrease in sea levels from increased snow outweighing the melting of Greenland's glaciers.

In another overview, all models clearly show both Greenland and Antarctica making small contributions over the century. However, Antarctica is generally soaking up more water than Greenland is shedding, as the IPCC predicts. The IPCC estimates that the very worst additional increase to be expected from Greenland could be eight inches over the century, but this is possible only in a model where CO_2 levels rise two to four times more than expected by 2100. Thus, there is very little support for the assumption of a twenty-foot sea rise.

In 2006, we finally got the longest temperature-data series for Greenland, and it shows why we might be hearing much more about melting, because temperatures there indeed increased dramatically in the 1990s. It seems that Greenland bypassed the general warming tendency since 1940 and instead cooled until the 1990s (not something we have heard a whole lot about). The temperature increases we see now only reach the levels of the 1920s and 1930s, and the increases happened even faster back then. The warmest year in Greenland is still 1941, and the two warmest decades were the 1930s and 1940s.

Antarctica's gigantic ice sheet began accumulating some thirty-five million years ago and has been a permanent fixture in the globe's environment ever since. The ice sheet is on average more than one mile thick and rises in many places to more than two miles. During the last ice age, the West Antarctic Ice Sheet was much larger; in the present, interglacial Antarctica has been adjusting to the warmer temperatures by losing ice.

Surprising to many, precipitation on Antarctica is so low that most of the continent is a desert (the world's largest, in fact). However, temperatures are cold enough (about $-29°F$ on average) that almost nothing melts or evaporates, and thus snow tends to accumulate.

However, much of the world's attention on Antarctica has focused on just a very small part, the Antarctic Peninsula, which stretches up to within six hundred miles of South America. Here it is warming, while the other 96 percent of Antarctica has cooled. The South Pole has seen its temperature decline since the beginning of measurements in 1957.

However, the Antarctic Peninsula has warmed dramatically—more than $3.6°F$ since the 1960s, several times the rate of global warming. In his film, Al Gore shows us how the ice is rapidly melting and how the less-than-poetically-named Larsen B ice shelf dissolved within thirty-five days in early 2002. The significance of this breakup relies on us believing that Larsen B has been intact since time immemorial, so that it now portends dramatically higher sea levels. But this is wrong.

Studies show that in the middle of our present interglacial age the Larsen area saw "widespread ice shelf breakup." It is likely that the Larsen area was open water from perhaps six thousand to two thousand years ago. The maximal ice shelf dates only from the Little Ice Age a couple of hundred years ago, and much of what has subsequently collapsed is of that vintage.

Moreover, the breakup of the ice shelf did not cause the sea level to rise, because it was already floating. While it probably led to ice shelves flowing more quickly into the sea and glaciers retreating at a faster pace, the story left out one important fact. The precipitation on the Antarctic Peninsula is increasing, probably due to climate change, and this likely outweighs the melting. That is, despite the spectacular pictures of Larsen B, the Antarctic Peninsula is probably participating in an overall *lowering* of sea levels.

This, too, is the story of the continent. While most of Antarctica is too cold for ice to begin to melt, more heat means more precipitation and consequently an increase in the Antarctic ice cover, or a *decrease* in sea levels in all mod-

PENGUINS IN DANGER?

Al Gore also shows us how the rising temperatures on the Antarctic Peninsula have dramatically affected the emperor penguins, who were the subject of the 2005 documentary *March of the Penguins*. This colony of penguins, over five hundred yards from the pioneering French research station Dumont d'Urville, has been constantly monitored since 1952. Its population was steady at around six thousand breeding pairs until the 1970s, when it dropped abruptly to about three thousand pairs, and it has remained stable since. This could possibly be linked to climate change, although the onetime decline makes it less likely. However, it is but a single, and rather small, colony out of about forty colonies around Antarctica. It is the best studied only because of its location. Some of the largest colonies contain more than twenty thousand pairs each, several of which may be increasing. The IUCN estimates that there are almost two hundred thousand pairs and that the population is stable, placing it in the category "least concern." Moreover, the other main Antarctic penguin, the Adélie, has in the same region seen an increase of more than 40 percent over the past twenty years, underscoring the problem in simply blaming global warming and not telling the full story.

els. While studies are uncertain as to whether Antarctica is accumulating or shedding ice right now, all models predict ever more net accumulation over this century.

Summing up, this does not mean we will not experience sea-level rise due to global warming, but it will not be twenty feet or more. It will be closer to one foot over the century—about the same as what we experienced over the past 150 years.

So what will be the consequences of this sea-level increase? Often, we are presented with a view of society passively accepting the ever rising seas. But that seems entirely unlikely, for we didn't sit passively over the past 150 years watching the waves lapping ever higher. Rational countries will choose strategies that will actually reduce costs while sea levels increase.

Let us look at some of the top models and the most likely future. Today, about ten million people get flooded each year in coastal zones, and about two hundred million are in danger of it. Even if there were no global warming, this number would increase, both because populations will increase and because coastal areas seem to be more attractive. This is evident in the United States, where the total population has quadrupled over the past century but the coastal population in Florida has increased more than fiftyfold.

Moreover, many countries and cities are also subsiding—basically sinking. Venice is perhaps the most famous one, but its sinking has many different causes. A more clear-cut example is Santa Clara, California, where near-continuous water usage from 1920 to the 1970s lowered the water table fifty-five yards and caused the valley floor to subside by more than four yards before water extraction was strongly regulated.

So what will happen by 2085? If we don't do anything, and even *without* global warming, we will see a sharp increase in the numbers of people flooded—up to twenty-five million annually, with the numbers in the danger zone reaching 450 million. As above, it seems rather unlikely that we would not take relatively cheap action through the use of barriers (like the Thames Barrier, which protects London from sea surges); dikes and levees; coastal protection; and—on rare occasions—giving up land. If we invest smartly, we will essentially have no people flooded by 2085, simply because we are richer and can afford greater protection.

With global warming, rising sea levels will mean many more people will get flooded—if we don't change. More than a foot of sea-level rise will cause about one hundred million people to get flooded each year. These are the numbers that you will often hear bandied about, but of course they entirely disregard the fact that societies will deal with the issue. If they survived over the past 150 years in relative poverty, it is likely that they will do so much more effectively when they are more affluent.

Indeed, with smart protection at fairly low cost it is likely that we will see not one hundred million people flooded each year but fewer than one million. Today we have ten million people getting flooded—eighty years from now, with higher sea levels and more global warming and many more people, we will not see this number increased tenfold but actually *decreased* more than tenfold, because of rich societies being able to deal with flooding much more effectively.

Actually, the results are even stronger. If we contemplate a more environmentally oriented future, in which sea-level rises would be lower, our insticts would be to expect fewer people flooded. However, such a future would also be a less rich one—the IPCC expects the average person in the standard future to make $72,700 in the 2080s, whereas a person in a more environmentally oriented (but less growth oriented) world would make only $50,600. Despite one-third less sea-level rise, the environmental world will likely see *more* people flooded, simply because it will be poorer and therefore less able to defend itself against rising waters.

We will lose very little dry land to sea-level rise. It is esti-mated that almost all nations in the world will establish near-maximal coastal protection, simply because it turns out that doing so is fairly cheap. This was also what we saw when we talked about the historic district of Miami Beach. For more than 180 of the world's 192 nations, coastal pro-tection will cost less than 0.1 percent of their GDP and approach total protection.

Even some of the vulnerable nations that are often used as poster children for the devastating effects of global warming will remain almost entirely protected. The most affected nation will be Micronesia, a federation of 607 small islands in the western Pacific with a total land area of only four times that of Washington, D.C. If nothing is done, Micronesia will likely lose some 21 percent of its area by the end of the century. With protection, it will lose just 0.18 per-cent of its land area. However, if we instead opt for an envi-ronmentally focused future of less sea-level rise and less economic growth, Micronesia will end up losing a *larger* land area—0.6 percent, more than three times as much. As above, this is because increases in wealth are more important than those in sea level—poorer nations will be less able to defend themselves against rising waters, even if they rise less quickly.

Also, notice that the most exposed nation is likely to retain more than 99.8 percent of its land area—not a veritable catastrophe. For Tuvalu, land loss with protection will be just 0.03 percent, whereas the environmental world will see lower incomes and a tripled loss. For the Maldives, the loss without protection would amount to 77 percent, but with protection the loss will be 0.0015 percent, with higher loss in the environmentally proactive world. For Vietnam, the loss will be about 0.02 percent of land area, and in Bangladesh the loss will—perhaps surprisingly—be virtually nil at 0.000034 percent. Again, in an environmental issue-dominated future, the losses would be much larger.

Why are these losses so much lower than what one usually hears? It is a simple matter of the costs and benefits facing each nation. Micronesia could lose 21 percent of its land at a cost of 12 percent of its GDP; however, for 7.4 percent of its GDP, it can save almost all its land, making protection the better deal. For all other nations, the deal is much better, and consequently the protection even higher. The 77 percent land loss for the Maldives is worth more than their entire GDP (122 percent)—whereas protection will cost about 0.04 percent of GDP, making almost every square foot worth saving. While Vietnam would lose some 15 percent of its land without protection and suffer an 8 percent GDP loss, its protection costs are about 0.04 percent of GDP. Thus, Vietnam will likewise save almost all of its area.

Another way to put this is to note that as sea levels have risen since 1850, we have allowed ourselves to lose very little land, exactly because its value was much higher than the cost of protecting it. Obviously, with richer nations and scarcer land, this relationship will hold throughout the coming century.

Yet a multitude of opinion makers tell us that the consequences will be disastrous and that we need to change our ways. Tony Blair tells us that "sea levels are rising and are forecast to rise another 88 centimeters [one yard] by 2100, threatening 100 million people globally who currently live below this level." Consequently, he says, we must implement Kyoto.

Yet, as with temperature, the models tell us that Kyoto will have a very small impact on sea-level rise. If everyone, including the United States and Australia, implemented Kyoto and stuck to the agreement throughout the century, it would—for significant costs—postpone sea-level rises about four years. Thus, the sea-level rise we would have expected to see in 2100 would first occur in 2104. Basically, it would leave the world poorer but not much better at dealing with the problems.

Likewise, Greenpeace tells us that the Maldives will be submerged and, "if current warming trends continue, cities like London, Bangkok and New York will end up below sea level—displacing millions and causing massive economic damage." The solution is rapid CO_2 reductions, and it "is our only hope to avoid disastrous sea-level rise." Yet this is far from what the best models tell us. The shrill messages of the Maldives being submerged clash with the actual estimates of 0.0015 percent dry-land loss. We will see very little dry-land lost, simply because it will pay off to take actions to avoid greater loss.

If our goal is to improve the welfare of people and the environment and not just reduce carbon emissions, we have to openly inquire how best to do so. Doing too little about climate change is definitely wrong, but so is doing too much. If we chose to focus on a world in which we and the developing world get richer, we will likely see vanishingly little land lost. It is only if our fear makes us go down the path of an environmentally oriented and much less rich future that our actions—although well intentioned—may cause more loss of land and a lower overall level of human welfare.

Extreme Weather, Extreme Hype

Stronger and more frequent hurricanes have become one of the standard exhibits of the global-warming concerns. The Natural Resources Defense Council tells us that "global warming doesn't create hurricanes, but it does make them stronger and more dangerous." The group Friends of the Earth proclaims, "Hurricanes in Florida. Storms in the UK. Extreme weather events are predicted to become more frequent because of climate change." Greenpeace tells us "there is strong evidence that extreme weather events— such as hurricanes—are increasing (and becoming more severe and frequent) because of climate change." The solution offered is invariably CO_2 cuts and adoption of Kyoto.

With the strong 2005 hurricane season and the devastation of New Orleans by Katrina, this message has reverberated even more powerfully. Al Gore spends twenty-six pages showing pictures of the suffering in New Orleans and names every single hurricane in 2005. Robert F. Kennedy Jr., when looking at the New Orleans tragedy, blamed it on the United States "derailing the Kyoto Protocol" and said that "now we are all learning what it's like to reap the whirlwind of fossil fuel dependence." Just a day after Katrina wrecked Louisiana, environmental commentator Ross Gelbspan told us that its "real name is global warming."

Yet these statements resemble the exaggerated stories of the polar bears. They are vastly overstated and often point us in the wrong direction.

So has global warming caused stronger and more frequent hurricanes? And can we expect more in the future? Let us here use the latest consensus statement from the UN World Meteorological Organization (parent organization of the IPCC), which is more recent and more specific but generally in agreement with the 2007 IPCC report. It makes three strong points.

"**1.** Though there is evidence both for and against the existence of a detectable anthropogenic [human-caused] signal in the tropical cyclone climate record to date, no firm conclusion can be made on this point."

That is, the strong statements of humans causing more and stronger hurricanes (or tropical cyclones, as researchers call them) are simply not well supported. We just don't know yet. Al Gore is incorrect when he tells us that there is a "scientific consensus that global warming is making hurricanes more powerful and more destructive."

"**2.** No individual tropical cyclone can be directly attributed to climate change."

The strong public statements on Hurricane Katrina are simply not supportable.

However, the third WMO consensus point is perhaps the most important one. We rarely care about hurricanes as such—what we care about is their damage. Do they kill people and cause widespread disruption? And with global warming, will they kill and disrupt even more? The answer is—perhaps surprisingly—that the whole hurricane debate is somewhat tangential to this important question.

"**3.** The recent increase in societal impact from tropical cyclones has largely been caused by rising concentrations of population and infrastructure in coastal regions."

While the theoretical debate over whether hurricanes are increasing or not increasing is unlikely to reach a clear-cut conclusion anytime soon, most observers end up pointing out how *damages* from hurricanes are rising dramatically and quickly. Gore ends his discussion on hurricanes by presenting the historically increasing hurricane and flooding costs, telling us it is the "unmistakable economic impact of global warming." He tells us that by 2040, weather-related disasters could cost as much as $1 trillion, "driven by climate change." The numbers are correct. But attributing them to global warming is wrong.

The global costs of climate-related disasters have indeed increased relentlessly over the past half century. Yet just comparing costs over long periods of time does not make sense without taking into account changes in population patterns and demography as well as economic growth. There are two and a half times as many people in the world today as there were in 1950; each of us is more than three times as rich; and we have used our wealth to move to scenic coastal areas—which are much more risk-prone.

Thus, there are many more people residing in much more vulnerable areas with many more assets to lose. In Florida, Dade and Broward counties are today home to more people than lived in 1930 in *all* 109 coastal counties from Texas to Virginia, along the Gulf and Atlantic coasts.

The U.S. hurricane damage over the past 105 years seems to confirm the general concern over global-warming impacts—that economic costs are increasing dramatically, with three recent years standing out: Hurricane Andrew in 1992, breaking all previous records; hurricanes Charley and Ivan setting a new record in 2004; only to be dwarfed by Katrina and the other hurricanes in 2005.

But a group of researchers began to wonder whether the reason why the early part of the century got off so much more cheaply was because there were fewer people and fewer assets to be harmed. Thus, they asked, What would the damage have been if all hurricanes throughout the last 105 years had hit the United States as it is today, with today's population and wealth? Suddenly, the picture changes dramatically. If the 1926 Great Miami Hurricane had hit today, it would have created the worst damage ever in U.S. hurricane history. Hitting just north of where Hurricane Andrew landed in 1992, this Category 4 hurricane would have plowed right into the Art Deco district and downtown Miami and cost $150 billion in damages— almost twice the damage of Katrina. As it hit in 1926, it did damage or destroy all downtown buildings, but as these were many fewer and much less valuable, the total damage was more than two hundred times less—$0.7 billion present-day dollars.

The same story goes for the second-most-damaging hurricane, from 1900, which hit Galveston head-on, inundating the whole of the city with eight to fifteen feet of water. Had it hit today, it would have caused damage of about $100 billion, but in 1900, with fewer and simpler buildings, it cost "only" some $0.6 billion.

When we compare all hurricanes to the damage they would have wrought today, Katrina ranks third at $81 billion, followed by the 1915 Galveston hurricane at $68 billion and then Hurricane Andrew at $56 billion. Thus, there is no real increase in damages over the century from hurricanes being more destructive. As the WMO consensus points out, what we are seeing is instead the effect of more people with more assets living closer to harm's way.

What this also tells us is that damages will continue to grow as long as more and more affluent people move closer to the sea. The Association of British Insurers found that had Hurricane Andrew hit in 2002 rather than 1992, "the losses would have been double, due to increased coastal development and rising asset values." According to one current insurance-industry report, "catastrophe losses should be expected to double roughly every 10 years because of increases in construction costs, increases in the number of structures and changes in their characteristics." This is why the UN expects $1 trillion in hurricane damages by 2040—not because it is "driven by climate change."

But how can we then help people who are potential victims of future Katrinas, Charleys, and Andrews? Research shows us that hurricane damages over the next half century are not best controlled by turning the big knobs of climate but rather by the relatively inexpensive social-policy knobs.

If society stays the same—no additional people living close to the coast, no additional costly and densely built neighborhoods—and the climate warms, causing the worst-case increase in hurricanes, the total effect by 2050 will be less than a 10 percent increase in damages. To put it differently, if we could stop the climatic factors right now, we would avoid 10 percent more damage by mid-century.

On the other hand, if climate stays the same—no more warming—but more people build more and more expensive buildings closer to the sea, as they have done in the past, we will see an increase of almost 500 percent in hurricane damages. To put it differently, if we could curb societal factors right now, we could prevent 500 percent more damage in fifty years' time.

So if we want to make a difference, which knob should we choose first—the one avoiding less than 10 percent damage increase or the one avoiding almost 500 percent damage increase? The difference in efficiency between the climate knob and the societal knob is more than a factor of fifty. This seems to suggest that policies addressing societal factors rather than climate policies will help much more and much faster. But of course, we have to ask *how much* we can reasonably turn the knobs.

With the climate knob, it is clear that just turning it toward Kyoto has been exceedingly difficult. If we managed to get everyone to participate in Kyoto and continue to adhere to its ever more binding reductions all the way to

2050, it would essentially mean a slightly lower temperature by then, entailing a decrease in hurricane damage of about half a percent.

What about the societal knob? We may be a bit hazy on how we could even go about turning it, simply because we hear very little about it. Clearly, we could try to stop people from moving to the sea or prevent them from building nicer houses, but that would be somewhat unrealistic and unwanted. But there are many other possibilities. We could better map vulnerability—some U.S. communities have never been evaluated for flood risk. This would abet evacuation plans, community education, and relief distribution. We could regulate vulnerable land, including the zoning, regulation, taxing, and public acquisition of land at risk. We could avoid state-subsidized, low-cost insurance that implicitly encourages people to settle in high-risk areas. We could improve building codes, so that new structures would be better able to withstand high winds, and we could better enforce existing codes. We could maintain and upgrade the protective infrastructure, such as dikes and levees. We could invest in improved forecasts, better warning systems, and more efficient evacuations. We could reduce environmental degradation, where loss of vegetation reduces soil's ability to absorb water and destabilizes slopes, leading to dangerous landslides. Likewise, we could protect wetlands and beaches that act as natural seawalls against hurricanes.

Let us look at just a single, simple example. After Hurricane Katrina, one insurance company found that the five hundred damaged locations that had implemented all the hurricane-loss prevention methods experienced only one-eighth the losses of those that had not done so. At a cost of $2.5 million, these building owners had avoided $500 million in losses. Often, simple structural measures, like bracing and securing roof trusses and walls using straps, clips, or adhesives, can yield big benefits.

With almost 90 percent total damage reductions from simple procedures, let us conservatively assume that we could at least cut loss increases in half through cheap and simple policy measures. This would, over the coming half century, mean a hurricane-damage reduction of 50 percent.

Thus, in a world with increasing hurricane damage from both global warming and societal factors, Kyoto could probably reduce the total increased damage by about 0.5 percent, while simple preventive measures could reduce that same damage by about 50 percent—one hundred times better. Moreover, Kyoto runs into trillions of dollars of cost, whereas the protective measures would be several orders of magnitude lower. Thus, if we care about helping potential victims of future Katrinas and Andrews, it appears unquestionable to suggest that we should focus on societal factors first.

In a nutshell, Katrina really does carry home the message. What caused the tragedy in New Orleans was not the hurricane itself—it was not a killer Category 5 but a regular Category 3 hurricane that had been predicted for years. Models and exercises had repeatedly shown that New Orleans was not ready for a direct hit. In the words of one expert, it "was a disaster waiting to happen," and, unlike with previous hits, New Orleans had just run out of luck.

One climatologist pointed out, "It is probable that this hurricane would have occurred irrespective of any recent increase in greenhouse gases." On the other hand, it was clear that the catastrophe happened because of bad planning, poorly maintained levees, and environmental degradation of the city's protective wetlands. So had you been in charge in the 1990s of helping potential victims of future hurricanes, you should not have worked to cut greenhouse gases first but instead invested heavily in better planning, better levees, and healthier wetlands. And that lesson remains as we try to prevent the losses from future Katrinas.

It is perhaps sobering to realize that while the industrialized world (and the big insurance companies) worry about increasing financial costs, hurricanes cost less in the third world but cause many more deaths. Yet here also the message stays the same: we know effective action is possible to reduce disaster losses even in the face of poverty and dense population. During the 2004 hurricane season, Haiti and the Dominican Republic, which share the island of Hispaniola, provided us a powerful lesson. Julia Taft of the United Nations Development Program explained: "In the Dominican Republic, which has invested in hurricane shelters and emergency evacuation networks, the death toll was fewer than ten, as compared to an estimated two thousand in Haiti. . . . Haitians were a hundred times more likely to die in an equivalent storm than Dominicans."

We have to ask how we can do the most good, or in the words of two of the top specialists in hurricane losses:

> Those who justify the need for greenhouse gas reductions by exploiting the mounting human and economic toll of natural disasters worldwide are either ill-informed or dishonest. . . . Prescrib-

ing emissions reductions to forestall the future effects of disasters is like telling someone who is sedentary, obese, and alcoholic that the best way to improve his health is to wear a seat belt.

Flooding Rivers

The story of river flooding is much the same as what we saw with hurricanes. Unusually severe floods in the 1990s and the early 2000s from St. Louis in the United States to Poland, Germany, France, Switzerland, Spain, and the United Kingdom have garnered renewed attention to the problem of flooding.

Very often, those commenting on these floods make an explicit link to climate change. After the severe flood of Prague and Dresden in 2002, British prime minister Tony Blair, French president Jacques Chirac, and German chancellor Gerhard Schröder all used the flood as a prime example of why we must commit to Kyoto. According to Schröder, this flood showed us that "climate change is no longer a skeptical prognosis, but a bitter reality. This challenge demands decisive action from us," which he understood as a requirement for "all states to ratify the Kyoto Protocol."

And yes, it is true that global warming eventually will increase precipitation, especially heavy rains. Models also show that this will lead to more flooding. There is also some evidence that increased rain is already occurring, although the IPCC has still not been able to link it to global warming. Thus, Schröder might have a point.

But there are two problems with Schröder's argument.

First, the increasing rain does not seem to be translated into increasing flooding in rivers. This holds true in a global sample of almost two hundred rivers, in which twenty-seven did indeed show increasing high flows, but even more (thirty-one) rivers were decreased, and the large majority showed no trend. This also holds true for the smaller number of rivers around the world where we have observations stretching very far back. Why is this?

When studying U.S. rivers, we can see increasing precipitation causing increasing stream flows, but if we check *when* the increase happens, it turns out that it happens mostly during the fall, when there is generally lower flow and little risk of flooding, whereas it happens rarely in the spring and with high flows. Likewise, in Europe, a study of the two major rivers, the Oder and the Elbe (which flooded Prague and Dresden in 2002), showed that over the past centuries summer floods show no trend and winter floods actually have *decreased*.

This is well correlated with the historical evidence, which shows much greater flood risks in the colder climates of the Little Ice Age. With much snow and a late thaw, ice jams typically blocked a swollen river, producing high water levels, followed by floods and bursting dikes. This pattern was the main cause of flooding on the lower Rhine during the Little Ice Age, with almost all the dike bursts in the Netherlands being due to ice jams. These floods have decreased sharply in the twentieth century, on account of warming. Likewise, an analysis of the river Werra in Germany shows that the highest flood risk was in the 1700s. Along the river Vltava in the Czech Republic, floods have decreased over the past century.

We seem to have a very selective memory of floods, thinking that our age is special. And in a sense it is, but perhaps not in the way we think. In general, casualties due to flooding have been declining in Europe, with large-scale loss of life seen in episodes preceding the nineteenth century, while the twentieth-century death tolls were significantly lower, and deaths in the 1990s were lower still. Flooding has been pervasive throughout our history. All but two of the fifty-six major floods that affected Florence since 1177 (to pick just one example) occurred before 1844.

What does set our period apart is that *economic losses* from floods have been rising sharply in recent years, constituting some 25 percent of all economic losses from natural disasters over the last fifty-five years. However, just as with hurricanes, this tendency seems to be much more related to factors other than climate change—and this is the second problem with Chancellor Schröder's point.

No matter what the climatic future holds, flood impacts may continue to get worse. The congressional Office of Technology Assessment pointed out that "vulnerability to flood damages is likely to continue to grow" mainly because populations in flood-prone areas continue to increase, putting more property and greater numbers of people at risk, while flood-moderating wetlands continue to be destroyed.

This is in no small way due to the widespread use of levees. The United States has about twenty-five thousand miles of levees—enough to encircle the world at the equator. The problem is that with levees people take fewer precautions ("We're safe behind the levee") and are more likely to build expensive structures behind them. This would be fine if flood risk could be reduced to zero, but, as the National Academy of Sciences points out, "it is short-sighted and foolish to regard even the most reliable levee system as fail-safe." Thus, losses are likely to be much higher when the levee system inevitably fails.

Moreover, levees themselves tend to increase flooding. Imagine a river one thousand yards across. Now assume that it is pinched to five hundred yards by a levee. In order for the river to pass at the same flow, it must rise to a higher elevation both at the pinch point and upstream. Likewise, levees can increase downstream flood levels by reducing the floodplain's ability to store water. In a place that histori-cally would have been flooded, a levee protecting the land also denies the river temporary reprieve, essentially passing the flood quicker and more massively downstream. Thus, whenever a levee is constructed, it means water levels will increase upstream and downstream, leading other loca-tions to increase their levees, resulting in an ill-fated game of leapfrog.

This increasing flooding has been systematically docu-mented for the lower Missouri River. Flows that were fully contained within the Missouri channel in the early twenti-eth century now create floods, and extreme high flows today are four yards *higher* than they would have been in the 1930s.

As with hurricanes, we have a situation where by far the larger part of the increasing damage comes from societal rather than climatic factors. In 1929, the average annual flood damage in the United States was about half a billion dollars (in present-day value). Today, the damage is closer to $5 billion—or ten times more. But of course, we also have many more people with more wealth living in or near floodplains.

One way to redress this would be to look at how much of the nation's goods get damaged each year in floods. This clearly changes the picture. In 1929, $200 of each $1 million worth of goods got damaged, whereas today only $70 of each $1 million is lost. This indicates that as society has more tangible wealth, while more goods will get damaged in floods, the damage will constitute a smaller and smaller proportion of the total wealth. Overall, floods are not getting more damaging but *less*.

This does not mean we should not strive to make floods even less damaging. But how do we do this best?

The only large-scale study of the comparative climatic and societal effects on flooding comes from the British government's Foresight studies. Here it is shown that by implementing Kyoto we might be able to achieve a 3 percent reduction in fluvial and coastal flooding damages. If, on the other hand, we focus on dealing with the concrete flooding issues, we have many opportunities to reverse the rising damage trend, stabilize it, and perhaps even make it decline. We would inform people better about the flood risks, leading to less overconfident placements and encouraging more precautions. This would also entail no public subsidies to settlements in floodplains—unlike what is currently happening around St. Louis. We would have more stringent public planning, as is the case in places like Denver, Boulder, Austin, Phoenix, and Charlotte, where limited encroachment and guided development decrease flooding.

We would use levees more sparingly and allow some floodplains to do just that—be flooded sometimes, thereby providing buffers for the remaining areas. We would return some areas to wetlands, which would both decrease flooding and improve environmental quality.

The costs for such policies would be orders of magnitude lower than Kyoto's trillions of dollars, and they would provide better results much more quickly. The Foresight studies show that the sensible increased flood management comes at minimal cost and achieves a reduction of damages of more than 91 percent.

It is worth realizing the difference in efficiency from Kyoto and flood management. Using the United Kingdom example, for about 0.01 percent of GDP you get a benefit from damage reduction of 0.12 percent of GDP—a benefit-to-cost ratio of 11. From Kyoto, at the cost of 0.5 percent of GDP, you get a benefit of 0.00009 percent of GDP. Or, expressing it in equal terms, a dollar spent on flood management will reduce flooding 1,300 times better than a dollar spent on Kyoto.

Flooding is not getting out of hand; the costs are declining compared to total wealth. It is not predominantly a signal of global warming or of increasing heavy rains. The financial consequences of flooding are rapidly increasing, because of increased population and wealth behind levees that both occasionally fail and increase flooding elsewhere.

We come back to Chancellor Schröder urging us to help the future flood victims in Dresden by focusing on Kyoto. This would take large sums of money and would do virtually no good in the short run, postponing problems toward the end of the century by five years. Essentially it is a promise to the citizens of Dresden that their increasing flood costs will increase slightly more slowly.

On the other hand, societal policies such as better information, more stringent building policies, fewer subsidies, more flood areas, and more wetlands would be able to reduce or even stabilize losses at much lower cost and much sooner. Shouldn't that be our first priority?

A New Ice Age over Europe

The concern about global warming stopping the Gulf Stream is probably—at least as of now—the only issue that has had its own Hollywood disaster movie and a Pentagon worst-case scenario. Yet it is also one of the least well-founded fears.

The Gulf Stream and its associated currents flow from the Gulf of Mexico up along the U.S. coast to Newfoundland. There it breaks up, with most currents flowing toward the Canary Islands, but the ones we will focus on continuing toward the United Kingdom and Europe and the Norwegian Sea. As water evaporates from the warm current, it becomes ever more salty, dense, and cold, until it sinks in the seas between Norway and Newfoundland and returns along the bottom of the ocean, completing the so-called North Atlantic thermohaline circulation, or simply the "Atlantic conveyor."

The Gulf Stream last shut down some 8,200 years ago, when the final glacial ice sheets in North America melted and a giant pool of freshwater built up around the area of the Great Lakes. One day, the ice dam broke and an unprecedented amount of freshwater flooded the North Atlantic and disrupted the sinking salty water from the Atlantic conveyor. This pushed Europe into a little ice age for almost one thousand years. Now there are concerns that it could happen again. Of course, there is no glacial ice sheet and giant freshwater pool around today, but possibly meltwater from Greenland could trigger such a phenomenon.

Yet the relevance of such a story crucially depends on the Greenland melt being on the same order of magnitude as the ancient freshwater pool—and it is not. Over the coming century, the IPCC expects Greenland to melt almost one thousand times less than what happened 8,200 years ago. A team of modelers looked at what would happen if Greenland melted at triple the rate expected by the IPCC—or, as they put it, at the "upper limit on possible melting rates." Although they see a reduction in the Gulf Stream, they find "its overall characteristic is not changed" and that "abrupt climate change initiated by Greenland ice sheet melting is not a realistic scenario for the 21st century."

Yet this has definitely not stopped the worry. The original concern over a Gulf Stream breakdown originated in the 1980s with geochemist Wallace Broecker, who put the theory forward one week before world leaders met in Japan to agree to the Kyoto Protocol. It was popularized in 1998 by William Calvin in his article "The Great Climate Flip-flop," which graced the cover of *The Atlantic Monthly*. There he told us how an abrupt cooling could be triggered by global warming, with chilly consequences: "Europe's climate could become more like Siberia. Because such a cooling would occur too quickly for us to make readjustments in agricultural productivity and supply, it would be a potentially civilization-shattering affair, likely to cause an unprecedented population crash."

This line of thought was picked up by a think tank at the

Pentagon and worked into a global scenario in 2003. In 2004, *Fortune* magazine revealed "the Pentagon's Weather Nightmare," the story of which flashed around the world with headlines such as "Now the Pentagon Tells Bush: Climate Change Will Destroy Us." The study sketches a "plausible" scenario, talking explicitly about a repeat of the 8,200-year-old Gulf Stream collapse. It envisions a tripling in temperature increases and bigger hurricanes and storms, which by 2007 make The Hague, in the Netherlands, uninhabitable. By 2010, the Gulf Stream collapses, and by 2020 "Europe's climate is more like Siberia's." As abrupt cooling reduces productivity, "aggressive wars are likely to be fought over food, water, and energy." By 2030, nuclear war seems to be a likely outcome.

It is not hard to see how these stories became an inspiration for the Hollywood blockbuster *The Day after Tomorrow*. The movie is an excuse for breathtaking special effects as Manhattan is buried in thirty-story snowdrifts and Asia is hit by killer grapefruit-sized hail. The British queen's helicopter is frozen in midflight, and Los Angeles is hit by multiple 250 mph tornadoes. Amid it all, a fearless paleoclimatologist played by Dennis Quaid straps on his snowshoes to trek from Washington, D.C., to New York City to rescue his son. The bad guy is the vice president, who bears a striking resemblance to Vice President Dick Cheney. The Cheney doppelgänger arrogantly dismisses the Kyoto Protocol—it's too expensive—and rejects concern about climate change as fearmongering. The scriptwriters save him from death to subject him to a mea culpa public address at the movie's climax, broadcast live on the Weather Channel: "We thought that we could affect the Earth's delicate systems without suffering the consequences. We were wrong. . . . I was wrong."

Yet the problem with these terrifying forecasts is that they make great special effects but little sense. Even if we ignore the fact that Greenland will not supply the disruptive amount of freshwater, a shutdown of the Gulf Stream would not turn Europe into Siberia. In the event 8,200 years ago, Europe probably cooled some 2.7°F. Model estimates project the same kind of drops from future disruptions. For comparison, the average temperature difference between Siberia and Europe is 23°F.

The MIT ocean physicist Carl Wunsch has pointed out in *Nature* that the ice-age scares are much overblown. In fact, "the only way to produce an ocean circulation without a Gulf Stream is either to turn off the wind system, or to stop the Earth's rotation, or both."

The latest Gulf Stream scare came in late 2005. Over the past forty-seven years, scientists have measured the Gulf Stream just five times, using a ship sailing across the 25°N latitude from Africa to the Bahamas. Their last measurement was in 2004. Suddenly, the researchers found that possibly 30 percent less water was now headed toward the far north than before.

The headlines were predictable. *National Geographic* told us " 'Mini Ice Age' May Be Coming Soon." *New Scientist* chose the news as one of its top stories from 2005, showing that "global warming may soon spiral out of control" and "plunge western Europe into freezing winters and threaten climate systems worldwide." *The Sydney Morning Herald* even saw a way to make this a global issue in "Scientists

Forecast Global Cold Snap," which included a reference to *The Day After Tomorrow.*

A few people thought this concern might be a bit premature. If the Gulf Stream were really slowing down, we should have seen temperatures drop 2 to 3°F, yet no such change had been seen. Carl Wunsch pointed out the obvious hype: "The story is appealing, but it is a very extreme interpretation of the data. It's like measuring temperatures in Hamburg on five random days and then concluding that the climate is getting warmer or colder."

In order to get a better grip on the Gulf Stream, scientists moored nineteen observation stations across the Atlantic in 2004, which have provided continuous data since then. In late 2006, at the first scientific meeting on the accumulated data, it became clear that there is no sign the Gulf Stream is slowing down. In *Science* the headline ran, "False Alarm: Atlantic Conveyor Belt Hasn't Slowed Down After All." In *New Scientist* the headline was "No New Ice Age for Western Europe." Unfortunately, it seems no other major news outlets found this important enough to pass on to their readers.

This is also why the IPCC, in its 2007 report, is very clear about the Gulf Stream: "None of the current models simulates an abrupt reduction or shut-down." The IPCC's models expect somewhere from no change to a Gulf Stream reduction of 50 percent over the coming century, but *no* models show a complete shutdown.

But isn't it still bad that the Gulf Stream may slow down up to 50 percent? Again, this depends on your perspective. If our goal is intrinsically to keep the Gulf Stream the same, yes, it is bad (as would be any other change, positive or negative and of any size). But if—more plausibly—our goal is to improve human and environmental quality, it might actually be an *advantage*. If the Gulf Stream weakens over the coming century, it will be while temperatures in general increase due to global warming. All the advanced models show that the net outcome is still warming over all land areas. The weakening of the Gulf Stream would mean *less* warming over Europe—presumably the goal of climate policy today.

As the IPCC points out, Europe will still warm, even if the Gulf Stream shuts down completely:

> Europe will still experience warming since CO_2 warming overwhelms the cooling associated with the Gulf Stream reduction. In consequence, catastrophic scenarios about the beginning of an ice age triggered by a shut-down of the Gulf Stream are mere speculations, and no climate model has produced such an outcome. In fact, the processes leading to an ice age are sufficiently well understood and completely different from those discussed here, that we can confidently exclude this scenario.

Malaria in Vermont

For the Climate Conference in Milan in 2003, the World Health Organization published a book that estimated increased temperatures since the 1970s to have caused 150,000 deaths in 2000. Green organizations, political parties, and pundits have incessantly repeated this figure. Not surprisingly, a headline like "Climate Change Death Toll Put at 150,000" sells a lot of newspapers.

Yet let us take a peek under the hood of this number. WHO produced a three-hundred-page book that contains a large number of interesting papers, but the actual ascription of the 150,000 deaths is in just one chapter. There, the authors set out to estimate the effects of cold and heat deaths, malnutrition, diarrhea, malaria, and floods.

Floods turn out to be a minimal factor, and as we saw above they are probably also more related to societal changes. Malaria and malnutrition we will look at in this and the next chapter. But a curious thing happened with cold and heat deaths. While the authors spent three pages talking about heat and cold deaths, when they aggregated the numbers they simply *left out* cold and heat deaths, leading to the total death toll of 153,000.

Now, if it was because cold and heat deaths were minor issues, this may have been understandable, but as we saw above, this does not seem to be the case. If we make a rough estimate of the lives lost and saved by the temperature increase since the 1970s of 0.65°F, we get about 620,000 avoided cold deaths and 130,000 extra heat deaths. This of course dramatically influences the total outcome: instead of 150,000 dying of global warming, there are actually almost 200,000 more people surviving each year. Again, it is important to stress that this does not mean we should just embrace global warming. First, we need to remember that there is more to global warming than deaths from temperature and disease. Second, because many of the diseases kill young people, whereas cold deaths generally afflict old people, there is a larger loss of years and potential. Nevertheless, this number of 150,000 deaths from climate change has had tremendous leverage, and unfortunately it has been based on only doing part of the math. Had all the relevant data been included, it would more likely have shown the opposite picture.

We see the same claims in many other areas of disease and global warming, most importantly with malaria, which each year infects half a billion people and kills more than one million. Former UN secretary-general Kofi Annan told us "a warmer world is one in which infectious diseases such as malaria and yellow fever will spread further and faster." The scare is raised that malaria could even come to the shores of developed countries. The cover story from *U.S. News & World Report* that predicted waterlogged Art Deco hotels in Miami Beach also expected that in the future "malaria could be a public health threat in Vermont."

As in most stories, there is at core some truth to the claim that cases of malaria will increase with temperature, but many of the public statements made on the connection between global warming and malaria have ventured far beyond the reasonable. The malaria parasite dies below 61°F, so if we run a model to determine how many people live within areas that will go from below 61°F to above 61°F over the coming century, we can determine that there will be about three hundred million more people "at risk of malaria" by the 2080s. However, this emphatically does not mean that three hundred million more will be infected by malaria, as we will see below—only that this is the upper limit. The same models find that in 2007 a bit more than 5.5 billion people are at risk of malaria—that is, 84 percent of the world's population. This is more than ten times the number of people infected each year. Clearly, many factors besides climate determine whether or not malaria is a serious risk.

Perhaps the most important factor is better appreciated when we realize that malaria was endemic in most of the developed world just fifty to one hundred years ago. When I first started reading about malaria in England in the Little Ice Age, or the last Dutch malaria epidemic in 1943–46, or the Tennessee River Valley in 1933, when 30 percent of the population was affected by malaria, I found it hard to believe that I hadn't previously heard about these incidences. But there is abundant evidence—from eight references to malaria in the works of Shakespeare to carefully reconstructed malaria death rates from forty-three counties in England and Wales from 1840 to 1910. Rather, it shows how quickly we forget and how quickly we lose our sense of context.

Up till the 1940s, malaria was endemic in thirty-six U.S. states, including Washington, Oregon, Idaho, Montana, North Dakota, Minnesota, Wisconsin, Iowa, Illinois, Michigan, Indiana, Ohio, New York, Pennsylvania, and New Jersey. Dr. William Currie, writing in 1811, began his book *A View of the Diseases Most Prevalent in the U.S.A.* by describing malaria: "A fever of an intermitting or remitting type or form is an endemic of America, and more or less epidemic every year during the autumnal season in all low and moist situations in every part of the continent."

The permanent secretary of the California State Board of Health noted in 1875 "that malarial fevers and consumption constitute the most prevalent forms of disease."

Experts estimated that for the nineteenth century and first part of the twentieth, "malaria has caused more ill health and loss of life in Florida than all other insect-borne diseases combined." In 1920, almost 2 percent of the U.S. population had malaria each year, and by the mid-1930s the United States still experienced more than four hundred thousand cases each year.

Only after the Second World War were the Centers for Disease Control established, with their first major task to combat malaria. (This is also why CDC headquarters lie in Atlanta, because most of the malaria disease was found in the Southeast.) From 1947 to 1949 more than 4.5 million American homes had been sprayed, and in 1951 malaria was considered eradicated. A long list of factors caused the final eradication, including public-health improvements, better nutrition, increased population in cities (where mosquitoes have a weaker foothold), better access to medication, large-scale draining, and finally the mosquito-control programs, along with home sprays.

Perhaps tellingly, a 1944 analysis of malaria in Mississippi showed that malaria deaths had been declining from 1,500 in 1916 to fewer than 400 in 1937. Since the number of dead still fluctuated from year to year, the analysis tried to see if it could be related to rainfall or temperature. No connection. However, because Mississippi so greatly depended on just one product—cotton—there was a strong variability in most people's income. Thus, we see a strong inverse relation between the number of dead and Mississippi income: when incomes were high, malaria deaths were low and vice versa.

When you think about it, it makes good sense. When you

have high income, you are better fed, you keep your home in better repair—such as in regard to screening—and if you get sick you can better afford medical services and medicine. During low-income years, you have less food and less stamina, your house falls into disrepair, and you don't get proper medical attention.

What the history of malaria in Europe and the United States shows us is that we eliminated malaria while the world warmed over the past century and a half. While temperature does have an impact on malaria, it is clearly not the main factor. What matters much more is a wide array of factors, from nutrition and health care, to draining and mosquito eradication, to income and availability of medical treatments.

This is also why claims of malaria returning to Europe or the United States seem unfounded. A recent European study found that malaria risk will increase 15 percent in the United Kingdom due to climate change. However, since there is an effective national health system in place, this essentially means 15 percent more of zero malaria. Thus, it concludes that "in Britain, a 15% rise in risk might have been important in the 19th century, but such a rise is now highly unlikely to lead to the reestablishment of indigenous malaria." Likewise for the United States, an otherwise very pessimistic summary of health risks found that reestablishment of malaria due to a warming climate "seems unlikely" as long as the current infrastructure and health-care systems are maintained. We will not have malaria in Vermont.

Today, a lack of infrastructure, effective health systems, and good public policy can be seen in sub-Saharan Africa. A belt runs through the middle of the continent where temperatures and precipitation make conditions ripe for malaria. Much of the region has weak, poor, and often corrupt governments that find it hard to implement, enforce, and pay for large-scale draining and spraying. Moreover, concerns from Western governments, nongovernmental organizations, and local populations make it hard to utilize DDT, which is still the most cost-effective insecticide against mosquitoes and, properly used, has negligible environmental impact. These populations are also characterized by poor health, with malnutrition and HIV interlinking with malaria to worsen all three. At the same time, poverty makes it harder to establish preventive measures and obtain effective drugs.

Over the past twenty years we have seen malaria death rates rising in sub-Saharan Africa, though the general death rate has been declining. At the same time, temperatures have in general been increasing, which has made it easy to claim that global warming is the culprit. There has been a substantial literature trying to ascertain the veracity of this claim, and in general it is clear that at best temperature is only one of many important factors. A recent World Bank review found that the main cause of increasing death rates is not increasing temperatures or increasing poverty but the fact that chloroquine—the mainstay of treatment over the past fifty years—is increasingly failing because the malaria parasite is becoming resistant to it. There are new and effective combination treatments based on artemisinin available, but unfortunately they are about ten times more expensive.

So what should we do? This depends on how much global warming really matters to malaria. One way to get an upper limit on the importance of global warming is to look at the projections of populations at risk. These are the models that show an additional three hundred million or so people living in areas that could harbor malaria in the 2080s because increasing temperatures will expand the area where the parasite can survive.

Again, it is important to point out that this is an absolute upper bound for several reasons. First, much of the increase will be in Europe and the United States, where strong health-care systems and infrastructure mean that malaria will not reestablish itself. Second, the analysis does not take into account better technology and higher incomes. Of course, when developing countries go from an average income per person of $5,000 today to $100,000 in 2100, it seems unrealistic to assume that this will not mean more protection and less malaria. Finally, the models also disregard that increasing urbanization will decrease the incidence of malaria.

This is also reflected in a careful analysis where researchers tried to see whether more people at risk would *actually* lead to more malaria. Their findings "showed remarkably few changes, even under the most extreme scenarios."

Nevertheless, let us try to see the worst-case scenario for what climate change will do to malaria. The same models that tell us three hundred million more people will be at malaria risk due to global warming also tell us what will happen *without* climate change. They project an increase from 4.4 billion in 1990 to 8.8 billion people at risk in 2085. The total population at risk will thus be 9.1 billion out of a population of 10.7 billion.

But notice the proportions: 8.8 billion will be at risk from malaria in 2085 due to social factors, whereas 0.3 billion extra will be at risk due to global warming. Thus, even if we could entirely stop global warming today (which we can't), we would change malaria risk in 2085 by only 3.2 percent. More realistically, adoption of the Kyoto Protocol would reduce malaria risk by 0.2 percent in 80 years. As a model team tells us, even with a stringent climate policy "there is little clear effect even by the 2080s."

Compare this to current expectations that we can cut malaria incidence to about half by 2015 for about $3 billion annually—or 2 percent of the cost of Kyoto. This was the fourth priority in the Copenhagen Consensus. Because we can do this within a decade, whereas climate policy will take half a century or more, the difference in actual people helped is even more dramatic. Till 2085, Kyoto will keep about seventy million people from getting infected by malaria (about 0.1 percent of all malaria infections). In comparison, a simple and cheap halving of malaria incidence by 2015 will keep more than twenty-eight billion people from suffering. This policy will do about four hundred times more good at one-fiftieth of the cost.

Malaria is poverty driven, and it is not surprising that the poor bear the brunt of the malaria burden. In Zambia, more than 70 percent of the poorest fifth of children are infected, whereas in the richest fifth fewer than 30 percent are. Likewise, poor households across twenty-two countries are less likely to have mosquito nets and much less likely to have insecticide-treated nets than rich households. Rural children—who are often also poorer—are less likely to use nets and less likely to get treatment. Rich children in general get more treatment for malaria.

In many ways, this is no different from conditions in Mississippi in the 1930s—if you were poor, you had a substantially higher risk of getting malaria, not getting treatment, and dying. But people no longer die from malaria in Mississippi, simply because the state and the surrounding society are now rich enough to afford to avoid it. Average annual personal income in Mississippi in 1930 was $202, or $1,974 in today's money—compare this to today's income of $24,925. Over the past seventy years, Mississippi became more than twelve times richer; this is the same development we expect from developing countries over this century.

In sub-Saharan Africa, per capita income is still lower than that in Mississippi in the 1930s: just $745. Studies show that when countries get to an income of about $3,100 per person, they can eradicate malaria, because their personal wealth will allow them to buy more protection and treatment, while their societies will be sufficiently able to provide general health care and environmental management, such as house spraying and mosquito eradication. Thus, even with very pessimistic assumptions about Africa's growth rate, it seems likely that the continent will cross the $3,100 threshold around 2080. With all other regions passing the threshold sooner, this will essentially mean that malaria will be eradicated late in this century (and even sooner with a successful and cheap vaccine).

Supposedly our goal is to diminish malaria and other infectious diseases. Malaria is a disease related strongly to economic development and weakly to changing climate. Malaria was endemic in most of Europe and the United States just a century ago. Yet coming out of the Little Ice

Age, these societies fought and won the battle over malaria, simply because of increasing wealth and good social policy. If we try to help developing countries deal with malaria, we can focus on easy paths that will cut infections by more than twenty-eight billion rather than the much more expensive path of Kyoto, which may help avoid seventy million infections.

The problem is that often the climate argument is virtually the only one offered. A recent Associated Press story spends 774 words on telling us how climate is making malaria worse in Kenya, with twelve words toward the end tersely telling us that "preventive programs, such as distribution of mosquito nets, can halt malaria's spread." WHO points out in very blunt language that the linking of disease to climate change reflects "a tendency to link all phenomena to popular scientific themes and this may also be true of malaria epidemics in Africa." However, WHO finds that the real reason for the reemergence of malaria in Kenya is not climate change but drug resistance, lack of control of mosquitoes, and a variety of policy problems.

More Heat Means More Starvation?

Having enough food is perhaps one of the most basic and important issues for many people in the world. It is also the first of the UN's eight so-called Millennium Development Goals for 2015—namely, halving the proportion of people who suffer from hunger.

Many people worry that climate change will dramatically undermine our future ability to feed ourselves. Stories of how global warming will "greatly increase the number of hungry people" and of how we are facing "catastrophe" with "whole regions becoming unsuitable for producing food" abound. Yes, global warming might slightly slow food production, but the claims are vastly overplayed and again—if our concern truly lies with food security and the world's hungry—lead us to focus on the wrong solutions.

To put the issue in context, food availability has increased dramatically over the past four decades. The average person in the developing world has experienced a 40 percent increase in available calories. Likewise, the proportion of malnourished has dropped from 50 percent to less than 17 percent. The UN expects these positive trends to continue at least till 2050 with another 20 percentage points' calorie increase and malnourished dropping below 3 percent. Yet 3 percent in 2050 will still be a massive 290 million people, and that should be cause for global concern. They will not be hungry because we cannot produce enough food. They will be hungry because they are poor and cannot create a demand for extra agricultural production.

A few large-scale surveys have looked at the effect of climate change on agricultural production together with the global food-trade system. There are four crucial findings generally shared among them.

First, there's good news. All models envision a large increase in agricultural output—more than a doubling of cereal production over the coming century. Thus, we will be able to feed the world ever better.

Second, the impact of global warming on production will probably be negative but in total very modest. For the most pessimistic models and the most pessimistic climate impacts, the total reduction compared to a scenario without any climate change is 1.4 percent. A lower climate impact and the most optimistic model actually forecast a net *increase* in agricultural production of 1.7 percent.

To put these numbers in perspective, the average annual growth rate for agriculture over the past thirty years was 1.7 percent. In the most negative scenario, the loss of 1.4 percent production over the coming century is less than one year of today's productivity increase. In other words, the total loss from climate change in the twenty-first century is equivalent of the world agricultural output doubling by, say, 2081 rather than 2080.

This will have very little impact on the global economy. The total agricultural GDP will probably see a change between a 1.5 percent reduction and a 2.6 percent increase by the 2080s. However, agriculture will constitute less than 1 percent of the total GDP; thus, the total economic impact will be minuscule, with at worst a 0.015 percent reduction.

Third, while globally there will be very little change, regionally this is not true. Global warming in general has a negative impact on third-world agriculture, whereas it has a positive impact on first-world farming. This is because temperature increases are good for farmers in high latitudes, where more warmth will lead to longer growing seasons, multiple harvests, and higher yields. For farmers in tropical countries—typically, third-world countries—higher temperatures mean lower agricultural productivity. For both places, however, CO_2 in itself counts as a positive factor, since it acts as a fertilizer, making crops grow more everywhere.

In worst-case scenarios, this will mean a 7 percent yield

decrease in the developing world and a 3 percent yield increase in the developed world. For some hard-hit places, this can mean relative yield reductions of 10 to 20 percent over the coming century. On the surface, this seems like a serious concern. However, this does not mean that these countries will see absolute production declines. Because of increased yields, better technology, and more farmland, these places are still likely to see production increases over the century by about 270 percent.

Over the coming century, there will be a growing dependence of developing nations on food imports from developed countries. However, this is not primarily a global-warming phenomenon but a consequence of more people and relatively less arable land in the developing world. Even without global warming, imports for the least-developed countries would double over the century because of demographics. Global warming causes these imports to increase by a further 10 to 40 percent.

However, we have to remember that in 2080 developing-country consumers are going to be much richer than they are today. One modeling team points out that future third-world consumers "are largely separated from agricultural production processes, dwelling in cities and earning incomes in the non-agricultural sectors. As in today's developed countries, consumption levels depend largely on food prices and incomes rather than on changes in domestic agricultural production."

Fourth, global warming will likely mean slightly more malnourished people, because food production will decline slightly. In the most likely scenario, it means increasing the number of malnourished in 2080 from 108 million to

136 million. (In other scenarios, global warming can actually lead to *fewer* malnourished.) It is important to put these numbers in context. As we saw above, the world now has about eight hundred million malnourished. Over the coming century, we will add at least three billion more people, yet it is likely that we will end up with *many fewer* malnourished—around 136 million (or 108 million without global warming).

However, how many hungry the world will end up with depends much less on climate than on demographics and income. The IPCC scenarios show that our political choices on global warming might cause 28 million more people to be malnourished, whereas our choices on economics and demographics would cause between 90 million and more than a billion more malnourished.

Once again, we have the situation where social-policy choices matter much more than climate choices. If we choose a society where we stop climate change, at best, we can reduce the number of malnourished some 28 million (but if we are unlucky, we could actually end up with *more* malnourished). If we choose a society with higher populations or with slower economic growth, we could end up with 975 million more malnourished. It is clear that we must find the scenario that offers the low absolute number of malnourished, which correlates with high income.

If this sounds like a recurrent point, it is. Even the modelers themselves point out that "what emerges from the [studies on malaria, hunger, water, and so on] is that these differences, flowing from different pathways of development, are frequently more important than climate change itself in influencing the scale and distribution of global and regional impacts."

Another way to see this is to realize that in a rich world the last 108 million or 136 million hungry are results of a *political* priority, one we could easily afford to reduce or eliminate. As pointed out by the modeling team: "To put it very bluntly, for the wealthy societies—even the currently poor regions are assumed to reach economic levels exceeding in per capita terms current average OECD [Organization for Economic Cooperation and Development] incomes—hunger is a marginal issue and remains so even with climate change."

Thus, using climate policy to obtain a small reduction in hunger is simply not the best strategy. If we implemented Kyoto, this would reduce malnutrition by two million people in 2080 for about $180 billion annually. However, if we really care about helping the hungry, we can do much better. We could focus on simple measures like investing in agriculture—improved soil health, water management, and agricultural technology research—and direct policies such as school meals and nutrient fortification (such as adding iodine to salt). The UN estimates that we could reduce hunger by 229 million people by 2015 for about $10 billion annually.

Again, it is worth pausing at these results: we can avoid 229 million people going hungry throughout this century for $10 billion annually. For the same amount of money spent on Kyoto, we can help one-eighteenth that number toward the end of the century. If we look at the effect over the entire century, it is the equivalent of avoiding just

39,000 malnourished annually. Each time our investment in climate saves one person from hunger, a similar investment in direct hunger policies could save more than five thousand people. This, of course, is why the Copenhagen Consensus results put malnutrition all the way up in second place (right after HIV), scrapping of agricultural subsidies third, and agricultural research fifth.

I believe it is important we face up to these facts. If we really care for the hungry of the world, should we not try to save five thousand before saving one?

Water Shortages

Al Gore tells us that the devastating drought and hunger just below the Sahara (in the Sahel) are not caused by nature, corruption, or mismanagement. Rather, the more we understand about global warming, "the more it looks as if we may be the real culprits." In Gore's view, CO_2 caused a significant part of the troubles that the Sahel has been experiencing.

Water is an important indicator for future well-being. It has also long been a subject of worry in environmental circles, where the argument is that we're approaching a crisis. Yet widespread claims of a "full-scale emergency" are misleading. True, there are *regional* and *logistic* problems with water. We need to get better at using it. But basically we have sufficient water. In its 2006 world water report, the UN summarizes that lack of water "is primarily driven by an inefficient supply of services rather than by water shortages." The World Water Council stated it even more clearly in its summary: "There is a water crisis today. But the crisis is not about having too little water to satisfy our needs. It is a crisis of managing water so badly that billions of people—and the environment—suffer badly."

The typical way to measure water problems is to use the so-called water-stress index. It essentially looks at how much water is available per person in a catchment and counts people with less than one thousand cubic meters per year (about seven hundred gallons per day) as highly water stressed. Such a limit is obviously not rigorous, since technology (such as drip irrigation in agriculture) and trade (importing grain instead of growing it) can dramatically lower the need for water.

When looking forward to the end of the century, it is clear that more people are going to be water stressed. However, this is simply a consequence of an increasing population—with more people sharing the same water, more will drop to the water-stressed level. This, however, ignores that increased economic ability will probably more than compensate for the lower amount of water available. After all, in an industrial society the vast majority will not be involved in water-intensive industries like farming, and thus these water-intensive industries can to a larger degree be placed where most water is available.

However, our question is how climate change will affect water stress. One of the largest models from the United Kingdom Fast Track Assessment has calculated this on the basis of populations and available water in the 1,300 major watersheds around the world. The remarkable result is that global warming actually *reduces* the number of people living in water-stressed areas, with less water stress in warmer scenarios than in colder ones.

Today, we have about two billion people in watersheds that are water stressed. Without global warming, this will increase to almost three billion by the end of the century. With global warming, it will be lower than today, at less than 1.7 billion. This is because a warmer world also means more precipitation—in general, models predict about 5 percent more precipitation by 2100. This does not mean that all regions will get more rain. Most of the planet will get wetter, such as southern and eastern Asia and parts of Africa, but some parts will get drier, especially the Mediterranean area, central and southern Africa, and the southern United States.

If we look at Africa, Al Gore is right that global warming will likely add twenty-eight million people to central Africa's water stressed by 2080. The same tendency is also true for southern and northern Africa, which will see fifteen million more water stressed. But then it also needs to be said that twenty-three million in western Africa will experience *less* water stress, and forty-four million in eastern Africa likewise. In total, with warming we have an Africa where twenty-four million *fewer* will be water stressed.

It is important to note that to be useful the extra precipitation will sometimes need to be stored for the dry season (as does part of the current precipitation). However, the upshot is that global warming will actually make fewer people in the future become water stressed.

While climate policies will not help make water more available (and may actually decrease access to it), there are other very beneficial and inexpensive ways to improve access to water and sanitation. This is still needed for the one billion people without access to clean drinking water and the two and a half billion without sanitation. We could bring basic water and sanitation to all of these people within a decade for about $4 billion annually.

This would avert almost one billion cases of diarrhea each year. It would also have an important but often overlooked effect on three billion people: access will save on average two hundred hours per person each year, time now spent walking to and waiting for water and toilets. The total monetary value of this benefit would be more than $200 billion. Providing clean drinking water and sanitation was the seventh priority on the Copenhagen Consensus list.

But hasn't global warming done something bad with respect to water? What about the claim that the Sahel has suffered during its prolonged drought in the 1980s and 1990s and that "we helped manufacture the suffering in Africa"? It is absolutely correct that the Sahel has suffered and that climate models, when fed the surrounding sea temperatures, can re-create the drought there. However, when two sets of researchers asked whether all the climate models could re-create the Sahel drought and the surrounding sea-level temperatures from CO_2 emissions, the resounding answer was no. Of nineteen models, only eight could generate some degree of drought, but even the best could not simulate its magnitude, beginning, or duration. Seven models even produced *excessive* rain. One research team concluded that "the drought conditions were likely of natural origin."

Global warming will mean more precipitation and more water availability for more people. The future water challenge lies not primarily in regulating global warming but in ensuring that three billion people can get access to clean drinking water and sanitation. This small policy change would be remarkably inexpensive at $4 billion annually and would bring huge health and quality-of-life benefits to half the world's population.

4

The Politics of Global Warming

Climate Policies Aren't Our Only Option

This debate is about our generational mission. What do we want to achieve over the next forty years?

Global warming is happening; the consequences are important and mostly negative. It will cause more heat deaths, an increase in sea level, possibly more intense hurricanes, and more flooding. It will give rise to more malaria, starvation, and poverty. It is therefore not surprising that a vast array of environmental organizations, pundits, and world leaders have concluded that we must act to fix global warming.

The problem with this analysis is that it overlooks a simple but important fact. Cutting CO_2—even substantially—will not matter much for the problems on this list. From polar bears to water scarcity, as we have seen, we can do relatively little with climate policies and a lot more with social policies.

If we claim that our concern lies with people dying from climate effects, as in the European heat wave in 2003, we have to ask ourselves why we are primarily thinking about implementing expensive CO_2 cuts, which at best leave future communities warming slightly less quickly, still causing ever more heat deaths. Moreover, as warming will indeed prevent even *more* cold deaths, we have to ask why we are thinking about an expensive policy that will actually leave *more* people dead.

Yet other social policies could allow us to both capture the benefits of global warming through reduced cold deaths and deal with the lower but increasing heat deaths through cities cooled by water, parks, and white surfaces and through better availability of air-conditioning and medical care. This would be orders of magnitude cheaper and would do much more good. Would we not rather that be our generational mission?

We care about low-lying islands and people being flooded from rising sea levels. Then why talk primarily about limiting carbon emissions, which will reduce sea-level rise but also leave islanders less well-off, ultimately leaving them with *more* lost dry land?

We profess our concern for increasing hurricane damage in the United States and for the terrible toll such damage exerts on third-world countries. But why would we want to focus on CO_2 cuts when, at best, we can reduce damage only by 0.5 percent? If we want to curtail hurricane damage, we could do much more through social policies, such as improved and enforced building codes, upgraded levees, and reduced subsidized insurance. Such policies could reduce damage by more than 50 percent at a fraction of the climate-policy cost.

This pattern goes for many other issues connected to

global warming. For each polar bear we save through Kyoto, we can save more than eight hundred bears through a cessation of hunting. Which should be our generational mission?

With Kyoto we can avoid about 140,000 malaria deaths over the century. At one-sixtieth the cost, we can tackle malaria directly and avoid eighty-five million deaths. For every time we save one person from malaria death through climate policies, the same money could have saved 36,000 people through better antimalaria policies. Which should be our first mission?

When focused on water scarcity, we see that global warming actually makes water more available. We found that climate change would improve access for a net 1.2 billion people, and Kyoto would actually make matters *worse*. Is that our good argument for global-warming policies?

Yes, global warming makes flooding more likely, but other policies are vastly superior in dealing with flooding, from no public subsidies for floodplain building to stricter planning, fewer levees, allowing floodplains to act as natural buffers, and increased wetlands. For every dollar spent on Kyoto, we could do 1,300 times more good through smarter social policies.

Take hunger. Yes, global warming will probably mean more malnourished, but again tackling hunger through climate policies is simply vastly inefficient. For each person saved from malnutrition through Kyoto, simple policies—like investing in agricultural research—could save five thousand people.

With all these choices, we have to ask over and over again, When choosing our generational mission, which policies should come first? We have become fascinated by the big knob of climate change and been sold the idea that if we can just turn this one knob, we can ameliorate most other problems in the world. Yet this is demonstratively false.

What we must come to terms with is that even though CO_2 causes global warming, cutting CO_2 simply doesn't matter much for most of the world's important issues. From polar bears to poverty, we can do immensely better with other policies. This does not mean doing nothing about global warming. It simply means realizing that early and massive carbon reductions will prove costly, hard, and politically divisive and likely will end up making fairly little difference for the climate and very little difference for society. Moreover, it will likely take our attention away from many other issues where we can do much more good for the world and its environment.

What We Should Do: Drastic Increase in R&D

The fundamental problem with today's climate approach is that ever stricter emissions controls—as in Kyoto and a possible, tighter Kyoto II—are likely to be unworkable. It is perhaps worth dispelling the myth that Kyoto's troubles are due only to the rejection of the treaty by a recalcitrant Bush administration. First of all, there has consistently been a majority in the U.S. Senate against an even weaker version of Kyoto. But perhaps more important, besides the United States and its fellow nonratifier, Australia, many of the participants in Kyoto—including Canada, Japan, Spain, Portugal, Greece, Ireland, Italy, New Zealand, Finland, Norway,

Austria, and my own nation, Denmark—are failing to meet the treaty's requirements for CO_2 reductions, and many have little or no prospect of doing so before the treaty expires in 2012. Had Bush simply joined the many nations that outwardly suggest compliance but actually show little intention to fulfill it, it would have become obvious that the treaty was never going to work.

Kyoto unfortunately has become the symbol of opposition to a United States seemingly uninterested in the opinions of the rest of the world. Thus, Kyoto has received political resuscitation without being seriously questioned for its efficiency or achievability. And this is the real issue: Kyoto is at the same time impossibly ambitious and yet environmentally inconsequential. It attempts to change century-old energy patterns in fifteen years, ending up costing a fortune and delivering almost nothing.

It is worth noting that the same problem besets the new EU pledge, the first real commitment since Kyoto in 1997. In March 2007, the EU members promised that they would unilaterally cut emissions to 20 percent below 1990 levels by 2020. This would mean a 25 percent cut of emissions from what they would otherwise have been in 2020. Yet the effect on temperature turns out to be smaller than Kyoto's, postponing warming by the end of the century by about two years. The cost would be about $90 billion per year from 2020. Thus, we see the same pattern as in the well-established Kyoto Protocol: fairly small impact at fairly high cost.

Regulating CO_2 emissions is simply very difficult. CO_2 emissions over the past half century tell their own clear story: they have increased inexorably—they even increased

11 percent under the Clinton-Gore administration. If we set emissions in 1990 to 100, emissions had already hit 109 when the Kyoto Protocol was signed in 1997. Had Kyoto been carried through, it would not have reversed this trend, only slowed it—expecting emissions in 2010 to hit only 133, rather than 142.

But even that rather modest target was hard to establish. Changing national energy systems takes a long time and has huge costs. Those countries that are on paths toward meeting their targets were well on their way to doing so without Kyoto. The twelve nations that have reduced emissions the most since 1990 are all former Eastern Bloc countries that have undergone radical economic downturns. Germany, which has also managed considerable redutions, has done so largely because it absorbed a former Eastern Bloc country, and Great Britain's reductions are principally the consequence of actions by Margaret Thatcher in the 1980s to break the British coal union and move that nation's energy system away from coal and toward gas—for reasons that were predominantly political and economic, not environmental. In 1997 when Kyoto was negotiated, German and British emissions levels were already both 9 percent below their 1990 levels.

But many other countries have had a much harder time reaching the targets, especially the United States, Australia, and Canada, which are all experiencing population-growth rates of more than 10 percent per decade, which naturally tends to increase national emissions. With the United States dropping out, Kyoto will now lead to only a slight reduction in emissions—about 0.4 percent, from 142.7 to 142.2, with 1990 levels being 100.

Kyoto is not the first time we made such commitments. At the Earth Summit in Rio de Janeiro in 1992, leaders promised to cut emissions back to 1990 levels by 2000. The OECD countries overshot their target by more than 12 percent. These attempts clearly show that cutting emissions has been politically (and economically) very hard to carry through.

And consider the future. The International Energy Agency expects that emissions will continue to soar, not the least from China, India, and the other developing countries. The OECD countries will increase their emissions some 20 percent by 2030, whereas the developing countries will more than double their emissions. Changing that will be very difficult and costly. This is why a Kyoto II with more stringent CO_2 cuts and more participation, from both the United States and developing countries, will be a very hard sell indeed. Most experts also expect that the results of a Kyoto II would be rather weak.

There is an economic, a political, and a technological problem with the Kyoto process of ever increasing CO_2-emissions cuts. The fundamental *economic* problem with both Kyoto and its stricter follow-ups is that all macroeconomic models show that they are poor investments.

This, however, easily spills into the larger *political* problem of Kyoto. First, it will get progressively harder to convince people that they should pay substantial sums for rather negligible environmental improvements a century away. Second, as costs mount they will undercut the willing cooperation, since some nations will try to ride free and others will claim to adopt the restrictions but give up during the process without obvious penalties. Third, high emissions restrictions and low achievements will erode the support for future treaties. Clearly, the failure of Kyoto to deliver any substantial restriction lowers the chance of a successful follow-up.

Arguably, the *technological* problem with the Kyoto process has the greatest impact. In the long term, global warming will be substantially reduced only if we can make a transition to a non-fossil-fuel economy. Many Kyoto advocates will claim that the Kyoto restrictions on emissions will spur new investments in research and development (R&D) that will enable us to move closer to such a transition. But this is backward. If we want technology, we should ask for and invest in technology. When we ask for immediate carbon-emissions reductions, we should not be surprised that the vast majority of the investments go toward that specific goal.

In the Kyoto treaty there is a complete absence of provisions to stimulate R&D. Not surprisingly, R&D in the important areas of global warming—renewable energy and energy efficiency—has dropped off since the early 1980s, and we see no increase as Kyoto kicks in. So although global warming desperately needs strong R&D investments in noncarbon or low-carbon energy technologies to control it, the Kyoto approach simply does not supply them.

At the end of the day, if cutting CO_2 costs twenty dollars per ton, the rich world might be willing to make some—if often symbolic—cuts at high price, but it is extremely unlikely that China, India, and the other developing countries will get on board. What we need to do to tackle climate change is to make this cost drop dramatically. If we could cut CO_2 for, say, two dollars per ton, it would be much easier to get everyone to cooperate on massive cuts.

This is why I suggest that a much more appropriate response to climate change would be a worldwide commitment to R&D for non-carbon-emitting energy technologies, aiming to lower the costs of future CO_2 cuts. We should steer away from a Kyoto II that would impose ever stricter standards at high economic costs with low benefits and fragile political commitments and instead address the fundamental issue of finding new low-carbon energy technologies that will carry us through the twenty-first century and beyond.

We should commit ourselves to spending 0.05 percent of GDP in such R&D of non-carbon-emitting energy technologies. This approach would cost about $25 billion per year. It would increase funding to R&D about ten times, yet it would be seven times cheaper than Kyoto and many more times cheaper than a Kyoto II. It could easily involve all nations, with richer nations automatically paying the larger share. It would let each country focus on its own vision of future energy needs, whether that means concentrating on renewable sources, nuclear energy, fusion, carbon storage, or conservation or searching for new and more exotic opportunities.

This money should be spent on research of all sorts, exploratory and applied; pilot programs to test and demonstrate promising new technologies; public-private partnerships to incentivize private-sector participation in high-risk ventures (just like those now used to get pharmaceutical companies to develop tropical-disease vaccines); training programs to expand the number of scientists and engineers working on a wide variety of energy R&D projects; government-procurement programs that can provide a predictable market for promising new technologies; prizes for the achievement of important technological thresholds; multilateral funds for collaborative international research; international research centers to help build a global innovation capacity (like the agricultural-research institutes that lay at the heart of the agricultural Green Revolution in the 1970s); and policy incentives to encourage adoption of existing and new energy-efficient technologies, which in turn foster incremental learning and innovation that often lead to rapidly improving performance and declining costs.

Preliminary studies indicate that such a level of R&D would be sufficient to stabilize CO_2 concentrations at twice the preindustrial levels, essentially meaning that such an investment should be able to limit temperatures to about a 4.5°F increase from today.

Such a massive global research effort would also have potentially huge innovation spin-offs, from energy-storage research giving us better cell-phone batteries to the truly unexpected finds that dramatically improve our world. The Apollo moon program, with a total price tag of about $200 billion, is perhaps the most well-known such investment for its spin-offs, ranging from computer miniaturization to CT and MRI scanners.

Because the costs are much lower and there will be many immediate innovation benefits, the political fragility of the process will disappear. The project will not be vulnerable to the occasional free rider, simply because most governments will pay fairly little and can appropriate a large share of the immediate benefits in patents and industry spin-offs. Countries will no longer have to be ever more strongly cajoled into ever more restrictive agreements. Rather, they will partake because it involves them in a low-cost, long-term, viable solution to global warming.

Thus, we need to abandon the Kyoto process, with its focus on costly but ineffective and politically fragile cuts in CO_2. Even stronger agreements would only marginally change the temperature. Even more ambitious cuts would do little to help the people who need it the most. And even if the process doesn't break down, the tenuous, convoluted, constant political battle takes our attention away from the many other issues where we can do so much more, more efficiently.

This is the real moral problem of the global-warming argument—it means well, but by almost expropriating the public agenda, trying to address the hardest problem, with the highest price tag and the least chance of success, it leaves little space, attention, and money for smarter and more realistic solutions.

Instead of Thinking, We Get Scared Witless

Global warming is being described in everyday media in ever more dire terms. In 2006, the Institute for Public Policy Research (IPPR) think tank, which is strongly in favor of CO_2 cuts, produced an analysis of the debate in the United Kingdom. It summarized the flavor thus:

> Climate change is most commonly constructed through the alarmist repertoire—as awesome, terrible, immense and beyond human control. This repertoire is seen everywhere and is used or

drawn on from across the ideological spectrum, in
broadsheets and tabloids, in popular magazines
and in campaign literature from government initiatives and environmental groups. It is typified by
an inflated or extreme lexicon, incorporating an
urgent tone and cinematic codes. It employs a
quasi-religious register of death and doom, and it
uses language of acceleration and irreversibility.

This kind of language makes any sensible policy dialogue
about our global choices impossible. In public debates, the
argument I hear most often is a variant of "If global warming is going to kill us all and lay waste to the world, this has
to be our top priority—everything else you talk about,
including HIV/AIDS, malnutrition, free trade, malaria, and
clean drinking water, may be noble but is utterly unimpor-
tant compared to global warming." Of course, if the deadly
description of global warming were correct, the inference
of its primacy would also be correct, but as we have seen,
global warming is nothing of the sort. It is one—but only
one—problem of many we will have to tackle through the
twenty-first century.

Yet these pervasive apocalyptic descriptions of global
warming persist, strongly aided by the media, which thrive
on bad news. And climate sells particularly well. The IPPR
points out that "alarmism might even become secretly
thrilling—effectively a form of 'climate porn.' "

Alarmism has a long history in the climate debate. Perhaps most chillingly, this was evident in the witch trials
in medieval Europe. After the Inquisition's eradication of
the actual heretics (like Cathars and Waldensians), most

witches from the early 1400s onward were accused of creating bad weather. The pope in 1484 recognized that witches "have blasted the produce of the earth, the grapes of the vine, the fruits of the trees, . . . vineyards, orchards, meadows, pasture-lands, corn, wheat and all other cereals." As Europe descended into the Little Ice Age, more and more areas experienced crop failure, high food prices, and hunger; witches became obvious scapegoats in weakly governed areas. As many as half a million individuals were executed between 1500 and 1700, and there was a strong correlation between low temperatures and high numbers of witchcraft trials across the European continent. Even today, such a climate link is still prevalent in sub-Saharan Africa, where extreme rainfall (both droughts and floods) is strongly linked to the killings of "witches"—in just one district in Tanzania, more than 170 women are killed each year.

Less violently, the wet summer of 1816 (caused by the eruption of the Indonesian volcano Tambora) was blamed by many in Europe on the new practice of using lightning conductors. The authorities had to issue grave warnings concerning violent and illegal acts against the conductors. Interestingly, the same conductors had some years earlier been blamed for widespread droughts. The wet summers of the 1910s and 1920s were blamed both on the extensive gunfire from the First World War and on the initiation of shortwave transatlantic radio.

In the early part of the 1900s, the world worried about a new ice age. In 1912, the *Los Angeles Times* told us the "Fifth Ice Age Is on the Way: Human Race Will Have to Fight for Its Existence against Cold." In 1923, the *Chicago Tribune* declared on its front page that a "Scientist Says Arctic Ice Will Wipe Out Canada" along with huge parts of Asia and Europe.

However, the world was already warming then, and papers began picking up that point in the 1930s, asking if it might be related to CO_2. In 1952, *The New York Times* reported that "the world has been getting warmer in the last half century." In 1959, it pointed out that glaciers were melting in Alaska and the "ice in the Arctic ocean is about half as thick as it was in the late nineteenth century." In 1969, it quoted a polar explorer as saying that "the Arctic pack ice is thinning and that the ocean at the North Pole may become an open sea within a decade or two."

However, by the 1940s, global mean temperatures had begun to fall, which by the 1970s led to claims that the Earth once more was heading toward a new ice age. One popular book described the world: "Between 1880 and 1950 Earth's climate was the warmest it has been in five thousand years. . . . It was a time of optimism. . . . The optimism has shriveled in the first chill of the cooling. Since the 1940s winters have become subtly longer, rains less dependable, storms more frequent throughout the world." Growing glaciers were now seen as a problem: "The rapid advance of some glaciers has threatened human settlements in Alaska, Iceland, Canada, China, and the Soviet Union." It was estimated that the cooling had already killed hundreds of thousands in the developing world and if proper measures were not taken would lead to "world famine, world chaos, and probably world war, and this could all come by the year 2000."

Science Digest pointed out in 1973 that "at this point, the world's climatologists are agreed on only two things: that we do not have the comfortable distance of tens of thousands of years to prepare for the next ice age, and that how carefully we monitor our atmospheric pollution will have direct bearing on the arrival and nature of this weather crisis. The sooner man confronts these facts, these scientists say, the safer he'll be. Once the freeze starts, it will be too late."

In 1975, the cover of the respected *Science News* had a picture of New York being engulfed by an encroaching glacier, with the words "The Ice Age Cometh" blazing across. The magazine told us that we may be approaching a full-blown ice age: "Again, this transition would involve only a small change in global temperature—two or three degrees—but the impact on civilization would be catastrophic." Other commentators worried that cooling would cause "worsening droughts." *The New York Times* ran a story headlined: "Scientists Ponder Why World's Climate Is Changing; A Major Cooling Widely Considered to Be Inevitable."

Of course, today there are much better arguments and more credible models underpinning our worry about global warming, and since our societies are adjusted to the present temperature, a major departure either way will entail costs. But notice how the descriptions typically talk only about the impending problems and conspicuously leave out any positive consequences. If we are worried about more malaria from warming today, a world that believes in cooling should appreciate the reduction of

infected areas. Equally, if we are worried about short growing seasons with a cooling world, we should be glad that
global warming will lengthen them.

Also notice how the descriptions tend toward hyperbole,
telling us how we might lose Canada and vast parts of
Europe and Asia to an advancing ice age or how "world
famine, world chaos, and probably world war" could come
by 2000. Many of these incorrect claims sound curiously
like the equally incorrect Pentagon claims today of an
impending ice age from a Gulf Stream shutdown where
"Europe's climate is more like Siberia's" and nuclear war
seems to be a likely outcome.

As a society, we use large amounts of resources to regulate all kinds of risks. If media attention to some of these
risks is disproportionate, we will end up overfocusing on
these issues and underfocusing on other issues where we
plausibly could do much more good. There is a real risk that
with global warming we are moving down such a path, out-
lining a false conflict between fossil fuels and human survival, ignoring a sensible dialogue on trade-offs.

This becomes clear when we observe how extreme global
warming is being described and put in a ludicrous context
by some of the leading participants. A collection of green
and development groups recently stated that development
could come to an end, with the world starting to backslide:
"After a decade of UN conferences designed to end poverty
and save the global environment, disasters—driven or
exacerbated by global warming—could spell out the end of
human development for the poor majority, and perilous
political and economic insecurity for the rest of the world."

Leading British commentator George Monbiot explained

that we need to find out "how to stop the planet burning" and said climate change is as devastating as nuclear war. The EU commissioner for the environment, Stavros Dimas, in early 2007 even claimed that we would need a "world war" on climate change. These statements ride on the back of incessant news coverage of bad weather being caused by global warming and new science predicting ever worse futures. *Time* tells us, "Be worried. Be *very* worried." *New Scientist* tells us that we are standing at "The Edge of the Abyss." Perhaps taking the accolades for most outrageous description, a popular magazine told us that with global warming some experts "predict a future in which our children see rainforests burst into flame and seas boil—unless we act today."

But the proverbial tide may be starting to turn toward a fair assessment of the issues. Some leading scientists are beginning to speak out against this one-sided alarm. One climate scientist has even wondered whether some of the dire predictions push the science too far: "Some of us are wondering if we have created a monster."

Crucially, one of the United Kingdom's top climate scientists began speaking out against this hysteria in late 2006. Mike Hulme is director of the Tyndall Centre for Climate Change Research. He said yes, climate change is real and humans are definitely partly responsible for it. But words like "catastrophic," phrases such as "climate change is worse than we thought," and claims that we are approaching "irreversible tipping in the Earth's climate" and are "at the point of no return" are simply used as "unguided weapons with which forlornly to threaten society into behavioural change."

Why is it not just campaigners, but politicians and scientists too, who are openly confusing the language of fear, terror and disaster with the observable physical reality of climate change, actively ignoring the careful hedging which surrounds science's predictions?

Hulme finds that in the desperation over the failure of Kyoto and in preparation for negotiations for a future treaty a stage-managing of the new language of catastrophe is taking place. Daringly, he also says catastrophespeak helps avoid cutbacks in climate science funding. But ultimately, says Hulme,

> we need to take a deep breath and pause. The language of catastrophe is not the language of science. It will not be visible in next year's global assessment from the world authority of the IPCC....
>
> Framing climate change as an issue which evokes fear and personal stress becomes a self-fulfilling prophecy. By "sexing it up" we exacerbate, through psychological amplifiers, the very risks we are trying to ward off.
>
> The careless (or conspiratorial?) translation of concern about Saddam Hussein's putative military threat into the case for WMD has had major geopolitical repercussions.
>
> We need to make sure the agents and agencies in our society which would seek to amplify climate change risks do not lead us down a similar counter-productive pathway.

Jumping on the bandwagon of catastrophe, sexing up the ramifications of global warming, and exploiting fears of disaster may be good for selling papers, captivating viewers, and getting attention. But its stark and unfounded scares cut us off from a sensible dialogue on the political and economic arguments for action here—and on the many other problems facing us now and in the future.

The Economics: The Loss of a Sensible Dialogue

Global warming will definitely not be costless. As we have seen, it will cause more heat deaths, it will increase sea levels, and it will plausibly cause more intense hurricanes and more flooding. It will give rise to more malaria, starvation, and poverty. This is an important message to convey.

Equally, doing something about global warming will not be costless. Switching from coal to gas or to renewables comes at a price. Restricting transportation will make the economy less efficient. Cutting back on hot showers, plane trips, and car use will leave you less well-off. It will also reduce the number of people being saved from cold, it will increase the number of water stressed, and it will allow fewer to get rich enough to avoid malaria, starvation, and poverty. This is also an important message to convey.

Doing something about global warming has both benefits and costs. How to weigh these correctly is clearly a discussion we need to have. But in the current environment of panic, the climate changes are portrayed as so severe and overwhelming that even talking about the costs simply seems inhumane, unreasonable, and uncaring. We just have to aim for the benefits and not think about the costs.

But whether we talk about the costs or not, someone still has to pay. Even if we don't debate our priorities, we still end up prioritizing. Even if we end up doing some good, we might easily have done much better. If we are to embark on the potentially most costly global-policy program ever, we might first want to be sure it is the best use of our resources.

Yet many commentators make light of the costs of climate change. When the costs of addressing climate change were assessed in the 2007 Intergovernmental Panel on Climate Change (IPCC) report at up to 3 percent of GDP by 2030, several commentators said this amount—more than $1.5 trillion annually, more than one and a half times the cost of the world's military expenditure—was "negligible." Simon Retallack, head of the climate change department at the Institute for Policy Research (IPPR), claimed, "We won't notice it."

Throughout his book, Al Gore makes no mention of the costs of seriously addressing global warming. However, his remarks elsewhere for replacing payroll taxes with a CO_2 tax entail a $140 per ton cost and a tax on gas of about $1.25 per gallon. In one respected model, the annual economic cost amounts to about $160 billion for the U.S. economy in 2015. This would cut U.S. emissions to about half of what they would have been by 2015 and about 25 percent by 2105. Yet since the United States will be responsible for an ever smaller amount of the total CO_2 emitted over the century, the total effect in 2100 will be a reduction of global temperature in the neighborhood of 0.2°F. Essentially, the proposal suggests that the United States carry through a Kyoto-type restriction all by itself.

Similarly, George Monbiot has written a book on

how every facet of global warming is calamitous. Climate change will simply destroy the conditions that make human life possible, he says. He concludes that climate change is the project we must put before all others, meaning cutting OECD member emissions by 96 percent come 2030, essentially shutting down our current fossil-fuel-driven economy. He is vague on the total costs but assures us they won't lead to an economic collapse. Yet he envisions that we will basically have to reorganize our planetary energy and transport systems in twenty-three years. It will mean the end of air travel and the beginning of an all-encompassing CO_2-rationing system, much like the one for energy that Britain had during the Second World War.

Again, it seems to me we need a dialogue on whether the benefits of radical action like this will outweigh the costs. The economists tell us absolutely not. However, Monbiot says that he will not participate in such a dialogue because "it is an amoral means of comparison." He maintains that when talking about the benefits, we cannot capture the suffering of the people harmed by Hurricane Katrina, cannot capture the value of those drowned, or indeed cannot capture the value of lost ecosystems or the climate itself.

This is a weak argument. The argument of Monbiot and most other drastic-carbon-cut advocates is exactly one long discourse trying to capture the value of everything—from people to ecosystems—that we should save. He makes the plea that since the impacts will be so phenomenally overwhelming, we should be willing to make phenomenal sacrifices. He tells us that we should throw our weight behind a generational challenge of cutting carbon emissions 96 percent in twenty-three years. But that weight, that effort, that generational investment, can then not be used in dealing with the world's many other challenges, such as HIV, malaria, malnutrition, or clean water. Monbiot truly makes his case for a prioritization.

Part of what Monbiot seems to dislike is the idea of calculating everything into one denomination—particularly a dollar one. I can understand that. It highlights the harsh consequences of our actions, and moreover the methodology is difficult. But if we are to make comparisons across many different and disparate areas, it is crucial that we maintain our objectivity, and the economic approach helps us to do so.

Instead, Monbiot and many others go for the simple rhetoric points. He mentions how it does not make sense to trade off air travel and its impacts: "Should a steward be sacrificed every time someone in Ethiopia dies of hunger?" This may have oratorical shock value but seems stubbornly intent on missing the point. Perhaps stopping flying is not the best way to keep Ethiopians from starving.

Since flying makes up about 3.5 percent of the climate impact today—and by 2050 will still make up only about 5 percent—even a complete halt to flying, at great cost, would do very little good for an Ethiopian. As the full Kyoto Protocol would cut the effective climate impact *more* than a total cessation of flying, and undoubtedly at much lower cost, Monbiot is essentially proposing we pay even more than Kyoto for even less than Kyoto's outcome. Monbiot is focusing on saving fewer than two million people from hunger, when efforts tens of times cheaper could help 229 million people much faster and much more.

According to Monbiot, insisting on comparing costs and benefits means "you have spent too much time with your calculator and not enough with human beings." Yet I would argue that if this approach means you end up helping five thousand Ethiopians each time Monbiot helps one, it certainly seems like you care more about humans, too.

Far from being amoral to compare costs and benefits, it is crucially moral to ask, How do we help the most? Can it really be moral to do anything less?

This is why the major peer-reviewed economic cost-benefit analyses show that climate change is real and that we should do something, but our cuts should be rather small. In the latest review, the previous research is summarized:

> These studies recommend that greenhouse gas emissions be reduced below business-as-usual forecasts, but the reductions suggested have been modest.

This was state-of-the-art economics until October 2006, when a six-hundred-page British government report under the direction of economist Sir Nicholas Stern came out and created headlines everywhere. The report presented how it saw the climate evidence, summed up rather nicely by *The New York Times:* it "predicted apocalyptic effects from climate change, including droughts, flooding, famine, skyrocketing malaria rates and the extinction of many animal species. These will happen during the current generation if changes are not made soon."

The report's two main economic points are actually rather straightforward. First, Stern finds that the overall costs and risks from climate change are equivalent to losing at least 5 percent of GDP now and forever, and possibly 20 percent. The report itself stresses that this is the equivalent of the great wars and the economic depression of the twentieth century. Second, strong action against global warming will cost just 1 percent of GDP.

Virtually everyone has come away with the understanding that Stern has made a cost-benefit analysis and shown that the benefit is 20 percent and the cost just 1 percent, making strong climate action a slam dunk. This has made Stern a very popular man. The British environment minister pointed out: "Nick Stern is now an international rock star in the climate change world."

Yet a raft of academic papers have now come out all strongly criticizing Stern, characterizing his report as a "political document" and liberally using words such as "substandard," "preposterous," "incompetent," "deeply flawed," and "neither balanced nor credible." While there is a long list of problems with the analysis, I think it is enough to point out three issues.

1. The review's presentation of the science is massively exaggerated toward scary scenarios. "The Review fails to present an accurate picture of scientific understanding of climate change issues," and its "analysis of the prospective impacts of possible global warming is consistently biased and selective—heavily tilted toward unwarranted alarm." Since the review clearly has its expertise in economics, it is disturbing that so many of its alarmist climate interpretations have been able to gain so much ground.

2. The damages from climate change (and the benefits of action) are vastly inflated. As several peer-reviewed papers point out, "the Stern Review does not present new data, or even a new model." How can it then find conclusions that are completely outside the standard range? It turns out that the review has counted damages several times and some-what arbitrarily increased the damages eightfold or more according to new and conjectural cost categories that have never been peer-reviewed. At the same time, the review changed a key parameter in all cost-benefit analyses to a value that gives huge damages. Oddly, it forgets to use this parameter for the costs below, where it would count against a strong policy response. The parameter is also vastly out of sync with our present-day behavior: it would suggest that we should today save 97.5 percent of our GDP for future generations. This is patently absurd—today's saving rate is about 15 percent in the United Kingdom.

3. The costs of action are vastly underestimated, continu-ing a well-known "appraisal optimism," which was also seen in the 1950s onward in very low cost estimates for nuclear power. It implausibly expects costs of renewables to drop sixfold by 2050. At the same time, Stern forgets to count any costs of action after 2050, although these costs escalate and will remain important far into the twenty-third century.

This means that the difference between the peer-reviewed literature and the non-peer-reviewed Stern report is massive. In the peer-reviewed studies, global-warming damages run at about 1 percent of GDP and costs at about 2 percent. It is important to say that you can't just compare the costs of the two, because incurring the costs doesn't avoid all the damage. But basically incurring a 2 percent cost for a 1 percent benefit is a bad deal, which explains why economic cost-benefit analyses recommend only moderate CO_2 reductions. However, Stern essentially turns the standard economic picture around without new evi-dence: damages way outside what the previous literature has found and costs much more optimistic. Moreover, Stern is not so careful as to point out that he really didn't do a cost-benefit study—clearly he should only have compared the costs to part of the damages avoided.

The most well-known climate economist, William Nordhaus, concludes that the Stern review is "a political document." *Nature* tells us that the British government has tried to recruit other researchers to undertake this study, presumably angling for the same politically convenient outcome. Mike Hulme points out that "this is not the last word of scientists and economists, it's the last word of civil servants."

The Stern review must be praised for having put the economics squarely back into the climate debate. Whether or not we like to acknowledge it, doing something about global warming will have both costs and benefits, and we need the dialogue on how much we should do. But the Stern review does not change the fact that all peer-reviewed economic analyses show we should reduce CO_2 emissions only moderately.

The Science: The Loss of a Sensible Dialogue

The rising stakes of global warming are also closing off dialogues in some fields of science. The IPCC has as its stated goal "to provide policy neutral information for decision-making." Yet its chairman, R. K. Pachauri, calls for immediate and very deep cuts in CO_2. This is clearly opting for one policy over others, such as lower cuts, adaptation, or doing nothing.

When climate scientists found serious errors in the 2001

report (the so-called hockey-stick graph showing temperatures stable throughout the last millennium until global warming kicked in 150 years ago), this was not seen as a science issue but immediately cast as a political problem. In the words of several climate scientists:

> When we recently established that the method behind the so-called "hockey-stick" curve of Northern Hemisphere temperature is flawed, this result was not so much attacked as scientifically flawed but was seen both in private conversations and public discourse as outright dangerous, because it could be instrumentalized and undermine the success of the IPCC process.

Respected—but skeptical—climate scientist Richard Lindzen from MIT points out that

> scientists who dissent from the alarmism have seen their grant funds disappear, their work derided, and themselves libeled as industry stooges, scientific hacks or worse. Consequently, lies about climate change gain credence even when they fly in the face of the science that supposedly is their basis.

Over the years it has become obvious that part of—and it is important to say only part of—the IPCC has become more politicized. Take the famous statement from the 2001 report that most warming in the last fifty years is due to humans. In April 2000, the text was supposed to read: "There has been a discernible human influence on global climate." The October 2000 draft stated: "It is likely that increasing concentrations of anthropogenic greenhouse gases have contributed substantially to the observed warming over the last 50 years." Yet in the official summary, the language was further toughened up to say that "most of the observed warming over the last 50 years is likely to have been due to the increase in greenhouse gas concentrations."

·   When asked about the scientific background for this change by *New Scientist,* the spokesman for the United Nations Environment Programme, Tim Higham, responded very honestly: "There was no new science, but the scientists wanted to present a clear and strong message to policy makers." When scientists—without new science—"sex up" their message, it is no longer just science. It is advancing a particular agenda—namely, that their area is more important for funding, attention, and rectification than it really is. Sending a stronger message to politicians is simply using science to play politics.

    Likewise, some parts of the IPCC report read more like an ecological vision statement. Climate policy is here used as a tool and justification for charting an alternative course of development that is seen by some as preferable. The report points out that because of environmental scarcity our cars and trains can't keep getting faster. But that is okay, since "it is doubtful that this trend really enhances the quality of life." Instead, the IPCC suggests that we should build cars and trains with lower top speeds and extol the qualities of sails on ships, biomass (which "has been the renewable resource base for humankind since time immemorial"), and bicycles. Likewise, it is suggested that in order to avoid demand for transport we should create regionalized economies.

The IPCC finds we need to reorient our individual lifestyles away from consumption. Because of climate change, we must focus on sharing resources (for example through coownership), choose free time instead of wealth, choose quality instead of quantity, and "increase freedom while containing consumption." The problem is that "the conditions of public acceptance of such options are not often present at the requisite large scale."

The IPCC then suggests that the reason why we are unwilling to accept slower (or no) cars or regionalized economies with bicycles but no international travel is that we have been indoctrinated by the media. We see TV characters as reference points for our own lives, shaping our values and identities. Consequently, the IPCC finds that the media could also help direct the path toward a more sustainable world: "Raising awareness among media professionals of the need for greenhouse gas mitigation and the role of the media in shaping lifestyles and aspirations could be an effective way to encourage a wider cultural shift."

Stanford University climatologist Stephen Schneider, one of the most publicly visible global-warming scientists, has accurately and surprisingly honestly considered the "ethical double bind" that might occur to a scientist who is also concerned about contributing to a better world, of which we have seen several expressions above. As a scientist, he focuses on truth; as a concerned citizen, he must take an interest in political efficacy. Quite obviously, Schneider finds that this presents a delicate dilemma, and he expresses the hope that one might be both honest and effective. However, as Schneider agonizes over this dilemma he does offer the following bit of unambiguous advice: "We need to get some broad base support, to capture the public's imagination. That, of course, entails getting loads of media coverage. So we have to offer up scary scenarios, make simplified, dramatic statements, and make little mention of any doubts we might have."

Such a strategy might be very politically effective, but it undermines the opportunity for society to make informed choices among different policy goals. Global warming is not the only issue, and having some scientists making scary scenarios and dramatic statements simply closes off the vital dialogue on social priorities.

The Politics: The Loss of a Sensible Dialogue

The increasing rhetoric of what Hulme calls "fear, terror and disaster" is also polarizing the political debate on global warming to the extent that it incapacitates any sensible dialogue. This is perhaps no more clear than in the growth of the phrase "climate change denier," which as of this writing now has more than 21,000 hits on Google. This phrase is typically used as a catchall for people who don't accept the standard interpretation that humanity is squarely to blame for global warming and that we should cut CO_2 emissions dramatically.

The semantic invocation of Holocaust deniers is often explicit and certainly represents a strong symbolic undercurrent. One Australian columnist has proposed outlawing climate-change denial: "David Irving is under arrest in Austria for Holocaust denial. Perhaps there is a case for making climate change denial an offence—it is a crime against humanity after all."

Mark Lynas is the author of a book revealing "the truth about our climate crisis." (Disclaimer: I first met Lynas when he threw a pie in my face in an Oxford bookstore as a media stunt for his then upcoming book.) He finds that climate denial is "in a similar moral category to Holocaust denial" and envisions Nuremberg-style "international criminal tribunals on those who will be partially but directly responsible for millions of deaths from starvation, famine and disease in decades ahead." Likewise, David Roberts from *Grist* talks about the "denial industry" and states that we should have "war crimes trials for these bastards—some sort of climate Nuremberg."

Even the top scientist of the IPCC, its chairman, R. K. Pachauri, has ventured into the Holocaust comparison. When presented with my economic analyses of doing the most good first, he compared my way of thinking to Hitler's: "Where is the difference between Lomborg's view on humans and Hitler's? You cannot just treat people like cattle."

Though the alternative view has not yet been outlawed and is not always Hitlerized, it is often ridiculed. Al Gore typically replies to critical questions thus: "Fifteen per cent of the population believe the Moon landing was actually staged in a movie lot in Arizona and somewhat fewer still believe the Earth is flat. I think they all get together with the global warming deniers on a Saturday night and party."

The problem here is that the debate gets closed off. When Gore was challenged on *The Oprah Winfrey Show* that his

ALTERNATE SOLUTIONS IGNORED

With all the fervor directed at global warming, you might imagine most people would be excited at the opportunity to deal with it more cheaply and more intelligently. But you would be wrong. It seems that only cutting CO_2 is good enough.

Over the past decades, a number of alternative solutions have been suggested. The reaction to each has been surprisingly similar; let's look at just one of the most recent.

Atmospheric physicist John Latham in 2006 suggested that we could increase the reflectivity of low-lying clouds by creating more salt droplets from the ocean. This augments a natural process (breaking waves are constantly throwing vast quantities of salt up into the atmosphere), and it carries little risk (since we could simply stop and the system would return to its normal state within days). Perhaps most important, it could potentially stabilize temperatures at today's levels—doing much, much better than Kyoto, at about 2 percent of its cost.

Yet the environmental groups seem curiously uninterested. Friends of the Earth says: "It's not something we think we should be spending money and time on." The group denies that it is being dismissive: "It's not a question of being dismissive; it's a question of whether this is worth any time and effort even thinking about." The same story with Greenpeace: "Greenpeace wouldn't be interested in this sort of thing. We're looking for reductions in the use of fossil fuels rather than these technologies that in all likelihood would come to nothing."

Although Latham's research has been published in *Nature*, the proposal might of course not work. But shouldn't we want to check out whether we could solve one of civilization's major problems at very low cost?

estimates of sea-level rise are unrealistically high and his claims about malaria in Nairobi are unsupported by facts, he simply responded that many of the organizations that come out with studies questioning the effects of global warming are funded by the worst polluters.

While I appreciate the underlying moral intent to do good for humanity, the unwavering certainty that CO_2 cuts are the best way to help is problematic, as we have seen. Glib comparisons to flat-earthers and Apollo deniers simply deflect considerations of real-world conditions and impacts in Africa. A sanctimonious cry for a climate Nuremberg—while ironic in that focusing on CO_2 cuts could likewise be said to divert attention from many other and better solutions—is simply an unwarranted cop-out from taking responsibility to debate policy options in democracies.

For many writers and world leaders, global warming has been seized upon as a subject that can lift them out of the tedious bickering of distributional politics and instead allow them to position themselves as humanitarians and states-men concerned with the grandest issue of the planet's survival. They can capture the high ground as defenders of the interests of humanity, distancing themselves from the everyday infighting of self-interested politics.

Global warming has for a long time been the perfect issue, because it allows the politician to talk about things that have grandeur and yet are close to people's hearts. It actually makes some taxes popular, and yet the true costs of policies are far removed. At a recent climate demonstration in London, protesters actually chanted: "What do we want? Carbon taxes! When do we want them? Now!"

Since the climate is constantly changing, there will always be a change that can be blamed on global warming, while it has an immediacy that communicates well with voters. One online editor has compiled a list of more than three hundred problems claimed in the popular press to be caused by global warming—from allergies, gender inequality, and maple-syrup shortages to yellow fever.

And perhaps most important, the real costs of cutting CO_2 are postponed, preferably to the next generation of politicians. Kyoto was negotiated in 1997, but the restrictions will first hit between 2008 and 2012. The politicians who could claim victory in 1997 will generally not be the same ones to bear the costs of abiding by the restrictions. Likewise, the California equivalent of Kyoto has reaped a lot of political goodwill for Governor Schwarzenegger since he signed it in September 2006, but the target time is a convenient fourteen years off, in 2020.

Many countries and the EU are beginning to suggest other long-term CO_2 cuts, where again the honor lies with the present-day promoters and the hard work with politicians far down the line. This is perhaps most evident with Tony Blair's proposal to reduce CO_2 emissions by 60 percent by 2050. This sounds grand—which undoubtedly was the intention—but is also very far in the future. Since 1997, British CO_2 emissions have *increased* by more than 3 percent. Thus, many in the United Kingdom were understandably skeptical of the new proposal, and more than half the parliamentarians suggested that there should be annual goals, which would suddenly mean evaluating the grand goals of 2050 by 2008. The Blair government was horrified and resisted this—at present, the solution seems to be five-year plans, which reasonably allow judgment to be put off.

Nevertheless, this shows that when the time comes to commit to the political rhetoric of global warming, support suddenly withers away, because governments know that CO_2 cuts will quickly become very expensive and likely be politically dangerous.

This then is the depressingly obvious but debilitating consequence of the many years of politicians, the media, and NGOs riding global warming, accepting and even reveling in the language of "fear, terror and disaster." We have created a situation that is portrayed as ever more apocalyptic, but we have lost the opportunity for a sensible dialogue.

If one suggests—as I do here—that we need to adopt a long-term perspective and that we should increase R&D in non-carbon-emitting energy technologies to $25 billion per year, most people's reaction is simply that this is nowhere near enough, as we are facing an imminent environmental Armageddon. When emotions run this high, people stop listening to evidence and instead suggest solutions that are ever grander but also ever more unrealistic.

We have institutionalized hypocrisy. Politicians will stoke the climate scares and claim that they will cut back CO_2 in fifteen to forty years, long after they have left office. But we haven't seen much in terms of actual cutbacks, because those would be tremendously politically damaging, not to mention the effects they would have on many people's daily lives. Perhaps even more clearly, politicians will talk about facing the greatest threat to humanity—CO_2—but still insist on opening new airports, as the British government has done repeatedly.

Likewise, the media will push the climate scare as the ultimate bad-news story, as often seen in *The Guardian* and *The Independent* in the United Kingdom. Yet at the same time, both papers carry travel offers to faraway destinations and ads for cars, cheap flights, and energy-intensive consumer products. If these papers took the global-warming threat seriously, they would stop accepting advertising for all the trappings of the "good life." The failure to do so illustrates both hypocrisy and our strong dependence on fossil fuels.

And we as voters are not blameless. We have let the politicians and the media inflate the climate scare and gone out chanting for higher carbon taxes. Yet when these taxes were actually considered, as in late 2006 in the United Kingdom, there was an outcry, because suddenly we couldn't just be green. We would actually have to pay.

We have to begin to be honest about two things. First, climate change is not an imminent planetary emergency that will bring down civilization. It is one, but only one, of many problems that we will have to deal with over this century and beyond. Second, there are no short-term fixes to this problem. In the words of two eminent climate economists:

> Stopping, or even significantly slowing, climate change will require deep emission cuts everywhere. This project will take 50 years at least, but probably a century or longer. The political will to support climate policy has to span across parties, continents, and generations.

If we are to get this support across parties, continents, and generations, we must discard the debilitating scares and re-create a sensible and unbiased dialogue about goals and means, about costs and benefits, in dealing with global warming and the world's other challenges.

5

Conclusion:
Making Our Top Priorities Cool

Time for a Sensible Dialogue

As we said at the outset, Al Gore is right: the debate about climate change is a debate about our generational mission. Essentially, in the next forty years, what do we want to have accomplished?

As we have seen, there is a lot of "fear, terror and disaster" being bandied about—a kind of choreographed screaming. We need to move to a more sensible and fact-based policy dialogue, in which we can hear the arguments, sensibly debate their merits, and find long-term solutions. Presumably, our goal is not just to cut carbon emissions but to do better for people and the environment.

Yes, climate change is a problem, but it is emphatically not the end of the world. Take sea-level rise. That sea levels will rise over the coming century about a foot—about as much as they rose over the past 150 years—is a problem, but not a catastrophe. Ask a very old person about the most important events that took place in the twentieth century. She will likely mention the two world wars, the cold war, and perhaps the IT revolution. But it is very unlikely that she will add, "Oh, and sea levels rose."

We dealt with sea levels rising in the past century, and we will do so in this century, too. This doesn't mean that the process will be unproblematic, but it is unhelpful—and incorrect—to posit it as the end of civilization.

Moreover, what we realized is that sea-level rise will be a much bigger problem for countries that are poor than for countries that are wealthy. In fact, if we work hard on reducing sea-level rises, it is likely that we will reduce the rise by 35 percent but at the same time make each person about 30 percent poorer. The upshot is that places such as Micronesia and Tuvalu will get three times *more* flooded, simply because the effects of lower incomes more than outweigh those of the reduced sea-level rise.

Thus, we cannot just talk about CO_2 when we talk about dealing with climate change. We need to bring into the dialogue considerations about *both* carbon emissions and economics, for the benefit of both humans and the environment.

Yes, we should take action on climate change, but we also need to be realistic. The United Kingdom is arguably the government with the most high-pitched rhetoric on the issue. Since the Labour government in 1997 promised to cut British emissions by a further 15 percent in 2010, emissions have *increased* 3 percent. Emissions even during the Clinton-Gore administration went up 11 percent.

Look at our past behavior. We promised to stabilize our emissions from 1990 to 2000 but increased them by 12 percent. We have promised to cut emissions in 2010 by 11 percent but will probably end up with a 0.7 percent cut. Many believe that dramatic political action will follow "if only people knew better and elected better politicians." But look at the facts. For the last ten years, there has been a dramatic difference between the U.S. and the EU citizenry in their awareness and concern over global warming—and likewise a dramatic difference between the Kyoto enthusiasm of the EU leadership and the Kyoto derision of the Bush administration. Yet the development in EU and U.S. emissions per person since 1990 is very similar: while the United States' have remained stable, EU emissions have actually *increased* 4 percent.

And even if the rich world managed to rein in its emissions, the vast majority of the twenty-first-century emissions will come from developing countries, as is evidenced by the dramatic increases (from low levels) in emissions from China and India.

In a surprisingly candid statement from Tony Blair at the Clinton Global Initiative in 2005, he said:

> I think if we are going to get action on this, we have got to start from the brutal honesty about the politics of how we deal with it. The truth is no country is going to cut its growth or consumption substantially in the light of a long-term environmental problem. What countries are prepared to do is to try to work together cooperatively to deal with this problem in a way that allows us to develop the science and technology in a beneficial way.

Similarly, one of the top economic researchers tells us: "Deep cuts in emissions will only be achieved if alternative energy technologies become available at reasonable prices."

We need to tone down the screaming and reclaim the sensible dialogue. **We should tax CO_2 at the economically correct level of about two dollars per ton, or maximally fourteen dollars per ton.** Yet let us not expect this will make any major difference. Such a tax would cut emissions by 5 percent and temperature by 0.16°F. But before we scoff at 5 percent, let us remember that the Kyoto Protocol, which caused us ten years of political and economic toil, will reduce emissions by just 0.4 percent in 2010.

However, this is not the solution. Neither such a small tax nor Kyoto nor the draconian proposals for future cuts will move us much toward finding better ways for the future. R&D in renewables and energy efficiency is at its lowest point in twenty-five years. We need to find a way that allows us to "develop the science and technology in a beneficial way," a way that will enable us to provide alternative energy technologies at reasonable prices.

This is why one of our generational challenges should be for **all nations to commit themselves to spending 0.05 percent of GDP in R&D of non-carbon-emitting energy technologies.** It would cost a relatively minor $25 billion per year (seven times cheaper than Kyoto and many times cheaper than a Kyoto II). It could embrace all nations and yet have the richer pay the larger share. It would let each country focus on its own vision of future energy needs, whether that means concentrating on renewable sources, nuclear energy, fusion, carbon storage, or conservation or searching for new and more exotic opportunities.

It would create a global research momentum that could

recapture the vision of delivering both a low-carbon and high-income world. It would carry a low price tag and high spin-off innovation benefits. It would also avoid ever stronger temptations to free riding and the ever more difficult negotiations over ever more restrictive Kyotos. It is plausible that it will enable us to stabilize climate at a reasonable level.

I believe this would be a way to bridge a century of parties, continents, and generations, creating a long-run, low-cost opportunity to create the alternative energy technologies that will power the future.

To move toward this goal, we need to start a sensible policy dialogue. This requires us to talk openly about priorities. Often there is a strong sentiment in any public discussion that we should do *anything* required to make a situation better. But clearly we don't actually do that. When we talk about schools, we know that larger numbers of teachers will likely give our kids better educations. Yet we do not pile on more and more teachers, simply because we also have to spend money on other issues. When we talk about hospitals, we know that access to better technical equipment is likely to provide better treatment, yet we don't supply an infinite amount of resources. When we talk about the environment, we know tougher restrictions will mean better protection, but they also have higher costs.

Look at a similar issue with respect to traffic fatalities. Most people don't realize that traffic deaths are one of the ten leading causes of death in the world. In the United States, 42,600 people die in traffic accidents and 2.8 million are injured each year. Globally, an estimated 1.2 million people die in traffic accidents, and 50 million are injured every year. Moreover, car traffic also has a strong effect on nature—cars kill about 57 million birds each year just in the United States.

About 2 percent of all deaths in the world come from traffic. About 90 percent of the traffic deaths occur in third-world countries. The total cost is a phenomenal $512 billion per year.

Due to increasing traffic especially in the third world and due to ever better overall health conditions, the World Health Organization estimates that by 2020 traffic fatalities will be the second leading cause of death in the world, right after heart disease.

Amazingly, we have the technology to make this go away. We would literally overnight save 1.2 million humans, eliminate $500 billion in damages, and stop hundreds of millions of birds from dying every year. Doing so would of course especially address the plight of the third world, which suffers by far the hardest from traffic deaths.

The answer is simply lowering speed limits to 5 mph. Nobody dies—likely we could avoid almost all of the 50 million injuries each year. But of course we will not do this. Why? The simple answer that almost all of us would offer is that the benefits from driving moderately fast vastly outweigh the costs. While the cost is obvious in terms of the killed and maimed, the benefits are much more prosaic and dispersed but nonetheless important—traffic interconnects our society by bringing goods at competitive prices to where we live, by bringing people together to where we work, and by letting us live where we like while remaining able to visit and meet with many others. A world moving at only 5 mph is a lot like a world gone medieval.

Notice this is not meant to be a flip example. We really could solve one of the world's top problems if we wanted to. We know traffic deaths are almost entirely caused by people, we have the technology to reduce them to zero, yet we seem to persist in exacerbating the problem each year, pushing traffic deaths to become the number-two killer in the world by 2020.

I would suggest that the comparison with global warming is insightful. We also know that global warming is strongly caused by people, and we have the technology to reduce it to zero, yet we seem to persist in going ahead and exacerbating the problem each year, causing the temperature to increase to new heights in 2020. Why? Because the benefits from moderately using fossil fuels vastly outweigh the costs. Yes, the costs are obvious in the "fear, terror and disaster" we read about in the papers every day, but the benefits, though much more prosaic, are nonetheless important. Fossil fuels give us low-cost light, heat, food, communication, and travel. Electrical air-conditioning means that people in the United States die far less from heat waves today than in the 1960s. Cheaper fuels would have avoided the deaths of a significant number of the 150,000 people who have died in the United Kingdom since 2000 due to cold winters. Fossil fuels allow food to be grown cheaply, and we can have access to fruits and vegetables year-round, which probably has reduced cancer rates by at least 25 percent. Cars give us an opportunity to get more space and nature around our homes and still keep our commuting time constant, whereas communication and cheap flights have given ever more people the opportunity to experience other cultures and forge friendships globally.

In the third world, access to fossil fuels is crucial. About 1.6 billion people don't have access to electricity, which seriously impedes development. Two and a half billion people use biomass such as wood, waste, and dung to cook and keep warm. For many Indian women, searching for wood costs three hours each day, as they sometimes walk more than six miles per day. It also causes excess deforestation. About 1.3 million people—mostly women and children— die each year due to heavy indoor-air pollution. A switch from biomass to fossil fuels would dramatically improve 2.5 billion lives; the cost of $1.5 billion annually would be greatly superseded by benefits of about $90 billion. For both the developed and the developing world, a world without fossil fuels in the short or medium term is a lot like a world gone medieval.

This does not mean that we should not talk about how to reduce the impact of traffic and global warming. Most countries have strict regulations on speed limits—if they didn't, fatalities would be much higher. Yet studies also show that lowering the average speed in western Europe by just 3 mph could reduce fatalities by 25 percent—about ten thousand fewer killed each year. Apparently, democracies around Europe are not willing to give up the extra benefits from faster driving to save that many people. This is a political priority setting, where there is more than one solution. Speed limits across Europe are 50 or 60 kph in cities, but they are not 40 or 70. We can have a discussion about increasing the speed limit a bit, allowing for faster travel, or lowering it a bit, allowing for fewer killed. But 5 kph or 250 kph are unlikely to succeed.

This is parallel to the debate we are having about global warming. We can realistically talk about a CO_2 tax of two dollars or even fourteen dollars. But suggesting $140, as Al Gore does, seems to be far outside the envelope. Suggesting 96 percent carbon reduction for the OECD countries by 2030 seems a bit like suggesting 5 mph in the traffic debate. It is technically doable, but it is just very unlikely to come about.

Moreover, in the traffic case, we realize that in the short and medium term we are unlikely to move the speed limit much, and so instead we spend much of our attention on how else to reduce fatalities. This is where air bags, seat belts, motorcycle helmets, better highway construction, speed bumps, bike paths, and so on show their value. They can allow us to hold a reasonable speed *while* substantially reducing death and injury. If our discussion were entirely focused on just the speed limit and polarized between 65 mph and 5 mph, we would likely ignore the many easy, low-cost, and substantial injury reductions we could make.

Likewise with global warming, the debate has often become so fixated on CO_2 cuts that it neglects what presumably is our primary objective: to improve the quality of life and the environment. In the battles over whether we should cut 4 percent or 96 percent, we might easily forget that in the short and medium term we can help people much better through alternative policies. We can cut diseases, malnutrition, and lack of access to clean drinking water and sanitation while improving the economy with much cheaper polices that will have much greater impact.

THE PRECAUTIONARY PRINCIPLE CUTS BOTH WAYS

Cost-benefit analyses show that only very moderate CO_2 reduction is warranted, simply because cutting CO_2 is expensive and will do little good, and even then only a long time from now.

Often people will then claim that we should do so anyway because of the so-called precautionary principle. Although as a legal principle it really says only that lack of full scientific certainty about global warming should not be used as an excuse for doing nothing, it is typically reformulated as commonsense adages like "do no harm." While the legal principle is clearly correct, the adage approach is much trickier.

It is normally argued that "do no harm" obviously means that we must make strong CO_2 cuts now—after all, better to incur some costs now than incur harming heat later. This is problematic in several ways. First, if strong action (say 25 percent cuts) is warranted, why not cut 50 percent or 75 percent—that would mean even less harmful heat. This naturally ends with only one outcome—namely, 100 percent—which is the least harmful of all. Yet few would defend this outcome, because the costs would be too high, and this points to the need for trade-offs.

Second, the talk about "harm" from warming tends to neglect the fact that costs borne now will also harm us. Essentially, spending money on cutting emissions means less money for schools, hospitals, and all the other good items on the social budget.

Thus, when applying the precautionary principle carefully, we have to face the trade-off between the costs of avoiding future harm and the costs of incurring harm right now. We are then back to discussing the cost-benefit trade-off, and the precautionary principle really hasn't added any new insight. In practice, it seems the principle is mainly used to focus on preventing the harm from global warming, ignoring the harm from emissions cuts. In that way, it simply becomes a way of legitimizing an already established stance.

This is perhaps most clear in looking at the precautionary approach to traffic injuries. Clearly, the intention to "do no harm" in the traffic arena means cutting speed limits to 5 mph. Clearly, we also need to consider the harm inflicted by having no reasonably fast interconnections in modern society. And we're back to cost-benefit trade-offs.

And in the long term, our goal should be to make a transition to a low-carbon future so inexpensive that our kids and grandkids will want to do it and that China and India will want to as well. This is why we need to focus on R&D to improve the future.

I hope that in forty years we will not have to tell our kids that we went for a long series of essentially unsuccessful command-and-control Kyotos that had little or no effect on the climate but left them poorer and less able to deal with problems of the future. I also hope we will not have to say that we focused monomaniacally on global warming, neglecting most or all of our other challenges.

In forty years, we should be able to look our kids in the eyes and say we've managed the first part of the century-long effort to tackle climate change by making low-carbon energy technologies much cheaper and much more accessible. We should be able to tell our kids that our decisions have left them a world better equipped to deal with the future: richer, better fed, healthier, and with a better environment.

The Coolest Options

Reestablishing a sensible and fact-based dialogue on climate will mean that we can start doing the smart things first.

- Dealing with global warming will take a century and need a political will spanning parties, continents, and generations. We need to be in for the long haul and find a cost-effective strategy that doesn't splinter from overarching ambitions.

- We should cut CO_2 by more than what the emasculated Kyoto will manage, but still only by 5 percent, moving to 10 percent by the end of the century.
- We should increase our R&D in low-carbon energy ten-fold. Using 0.05 percent of GDP, or $25 billion annually, would enable us to stabilize climate at a reasonable level.

But we should also realize that how good we make the twenty-first century is not primarily a function of how we tackle global warming. If we look at the IPCC's economic estimates, even if we could magically make global warming go away we could maximally enrich this century by some $14.5 trillion. More realistically, global-warming policies can only do about $0.6 trillion worth of good, and if we overdo it we easily risk making the cure worse than the initial problem.

Contrast this with the IPCC's economic estimates. Which social path we choose—focusing on the economy or the environment, a world with higher or lower growth—is likely to have a much bigger impact of at least $553 trillion. If we focus too much on global warming we could easily end up making future generations far worse off, with the average person in the developing world missing out on incomes 70 percent higher in 2100.

With both attention and money in scarce supply, what matters is that we tackle the problems with the best solutions first, doing the most good throughout the century. What matters is making sure we leave our descendants with the most opportunities.

But perhaps just talking about trillions of dollars

seems a little insensitive. Let us instead talk about what we can do.

With a language of "fear, terror and disaster," climate change has captured most of our attention. But I say we need to cool it. We want to help the world—great. But it is not a given that the best way to help is to cut CO_2. What we have seen is that there are many other places where we could do much more good more quickly and more cheaply. An effort like Kyoto—which has spent much of the world's political will for the last decade and which will cost $180 billion annually if concluded in full yet provide surprisingly little benefit by the end of the century—makes the case clearly.

In table 2, I have tried to compare the efficiencies of Kyoto and a collection of smart policies discussed throughout this book, including the top priorities from the Copenhagen Consensus—let us call them "feel good" and "do good" strategies, respectively. Of course, Kyoto could be tweaked and made better, but it is what is on the table, and the differences are so huge and obvious that a better carbon-cuts proposal would only marginally change the table—and certainly not the outcome.

	FEEL GOOD (E.G., KYOTO)	DO GOOD
POLAR BEARS	0.06 saved	49 saved
TEMPERATURE DEATHS	84,000 *more* deaths	Already better in
FLOODING	$45 million in damage reduction	$60 billion in damage reduction ($5b)
HURRICANES	0.6% damage reduction	250% damage reduction ($5b)
MALARIA	70 million infections avoided	28 billion infections avoided ($3b)
POVERTY	1 million fewer	1 billion fewer
STARVATION	2 million fewer	229 million fewer
WATER STRESS	84 million *more*	Already better
HIV / AIDS		3.5 million lives saved ($7b)
MICRONUTRIENTS		Avert 1 billion-plus malnourished ($3b)
FREE TRADE		An extra $2.4 trillion annually
DRINKING WATER AND SANITATION		Give 3 billion access ($4b)
DEAL WITH CLIMATE EFFECTIVELY		$2/ton CO_2 tax
		R&D for low-carbon energy ($25b)
PRICE TAG	$180 billion / year	$52 billion / year

TABLE 2 The annual cost and efficiency of enacting Kyoto versus a collection of smart strategies (costs in parentheses).

We can debate just exactly what should be in the "do good" column. And we should. When we establish a sensible, fact-driven dialogue, we should debate whether we should spend more against hurricanes, less against flooding, more against malaria, or perhaps less against starvation. But overall it is obvious that the "do good" column is simply and overwhelmingly much better.

Many will say we should do everything in both columns. Yes, in principle we should do all good things, but until we do them all we should focus on doing the best things first. So far we have done pretty badly in both columns, so let's start with the smart one.

We will have to start seriously addressing down-to-earth but difficult subjects like the necessity of agricultural reform, wetlands restoration to minimize flood risk, the scrapping of subsidized insurance in hurricane zones, and better availability of medical care and air-conditioning. These conversations will have a pedestrian feel compared to the exciting charge of discussing climate change and global impacts over the coming centuries. But they will do amazing good for real people and real nature, with realistic time frames and plausible funding.

Over the next forty years what should we have accomplished?

I hope we will cool our conversation, rein in the exaggerations, and start focusing on where we can do the most good. This does not mean doing nothing about climate change, but it does mean having an open dialogue about its effects and solutions, a conversation about what our priorities should be.

When people spend five dollars to offset a ton of CO_2,

they do some good (probably providing about two dollars' worth of benefits to the world). But the same five dollars donated to a different organization could have done two hundred dollars' worth of social good if used for HIV/AIDS prevention or $150 worth of social good if used against malnutrition. I would like it to be cool to do two hundred dollars' worth of good before two dollars'.

I hope we will all work to make the best solutions cool. I would like to see college drives for mosquito nets against malaria before drives for adopting Kyoto. I would like to see demonstrations for improved building codes for hurricane resistance and improved enforcement before demonstrations against single-occupancy cars. I would like to see ardent letters to the editor about the unconscionable subsidies to industrialized agriculture keeping third-world farmers from achieving their potential before the ardent pleas to impose high carbon taxes. I would like it to be cool to be impassioned about doing the best first for the planet.

I hope we can look the coming generations squarely in the eyes and say that we didn't just do what seemed fashionably good; we massively and thoroughly changed the world for the better through simple, tested, and cool strategies. We didn't just do something that made us feel good; we did something that actually did good.

Acknowledgments

It has been a great privilege to be inspired by and to discuss, debate, and challenge the issues around global warming with the many people I have met and whose work I have read over the years. There is no way I can thank all of them—opponents as well as proponents.

I do, however, want to extend a strong thanks to all the scientists in the field, from climate to economics, from universities and research institutes, who actually measure and model the world in so many different ways and assemble and publish the bits and pieces of the information that are presented in this book. We could have no sensible debate without their work.

I also want to thank the many climate scientists and social scientists who have read part or all of the book, given me valuable inspiration, and clarified my thinking in numerous places. For various reasons, many did not want to be thanked. And of course, the customary caveat holds—only I am responsible for the content of this book.

I do want to say thank you to Henrik Meyer for giving me smart, continuous feedback: to Ulrik Larsen for improving many of my metaphors; to Egil Boisen for great input and for suggesting the title *Cool It;* to Richard Tol for making many of the economic arguments work better; to Roger Pielke for many great suggestions for this book throughout the time I've known him; to David Young for sharpening my arguments; to Chris Harrison for reminding me of all the other angles; and to my mentor Jørgen Poulsen for constantly reminding me of the bigger picture. Also thanks to my great colleagues at the Copenhagen Consensus Center: Tommy Petersen, Clemen Rasmussen, Elsebeth Søndergaard, Sonja Thomsen, Tobias Bang, and Maria Jakobsen.

I have been blessed with many great people at Knopf, not the least my editor, Marty Asher, who has consistently pressed me to refine my writing, along with Edward Kastenmeier, Zachary Wagman, and Arianna Cassidy. Also thanks to my copy editor, Timothy Mennel, who actually made me write good English. And finally a grateful nod to the talented publicity staff at Knopf: Paul Bogaards, Erinn Hartman, and Christina Malach. Thanks to Jeff Scott who first pushed me to write the book, and thanks to my agent, John Brockman, and his entire smart staff for believing in the book from day one.

This is a short book on a complex issue. But if we are to make our democracies count, finding the best generational mission, it is important that the information gets spread far and wide. If you feel you need more information, I'm also publishing a longer version of *Cool It*, with plenty of graphs and more explanation, with Cyan in the United Kingdom.

While every effort naturally has been made to ensure that all the information in this book is correct, errors will undoubtedly have crept in. I'll endeavor to post any mistakes to my website, www.lomborg.com.

Global warming is one of many problems facing us in the twenty-first century. I hope that this book can make us better able to fix our priorities and help the future do the best it can.

Bjorn Lomborg
COPENHAGEN, MAY 2007

Notes

CHAPTER 1

3 **The European Union calls it "one of:** (EU, 2001:208).

3 **prime minister Tony Blair of the United Kingdom sees it as:** (Blair, 2004b; Cowell, 2007).

3 **German chancellor Angela Merkel has vowed to:** (DW staff, 2006; Prodi, 2004).

3 **Presidential contenders from John McCain to Hillary:** (Buncombe, 2005).

3 **Several coalitions of states have set up:** (AP, 2006a; Pew Research Center, 2006).

3 **And of course, Al Gore has presented:** (Gore & Melcher Media, 2006).

4 **A raft of books titles warn:** (Gelbspan, 2004; Cox, 2005; Pearce, 2006).

4 **Pundits aiming to surpass one another even:** (Bunting, 2006).

4 **In 2006, *Time* did a special:** (*Time*, 2006).

4 ***Time* told us that due to global:** "And with sea ice vanishing, polar bears—prodigious swimmers but not inexhaustible ones—are starting to turn up drowned. 'There will be no polar ice by 2060,' says Larry Schweiger, president of the National Wildlife Federation. 'Somewhere along that path, the polar bear drops out' " (Kluger, 2006).

5 **Al Gore shows a picture similar:** (Gore & Melcher Media, 2006:146); see also (Iredale, 2005).

5 **The World Wildlife Fund actually warns that:** (Eilperin, 2004).

5 **In their pithy statement, "polar:** (BBC Anon., 2005).

5 ***The Independent* tells us:** A 2°C rise is now unavoidable, and "it will mean polar bears are wiped out in their Arctic homeland. The only place they can be seen is in a zoo" (McCarthy, 2006).

5 **Over the past few years:** (Berner et al., 2005; Hassol, 2004; Norris, Rosentrater, & Eid, 2002).

5 **Both relied extensively on research published in:** The World Conser-

vation Union is also known as the IUCN; the Polar Bear Specialist Group website is http://pbsg.npolar.no/default.htm (IUCN Species Survival Commission, 2001).

5 **But what this group really:** The IUCN counts twenty groups, but most commentators mention about nineteen subpopulations (IUCN Species Survival Commission, 2001:22).

5 **Moreover, it is reported that the:** (Krauss, 2006).

5 **Contrary to what you might expect—and:** (Michaels, 2004). See springtime temperatures at (Przybylak, 2000:606).

6 **Actually, there was a single sighting of:** (Monnett, Gleason, & Rotterman, 2005).

6 **That its population:** (Harden, 2005; WWF, 2006).

6 **Not mentioned, though:** (Stirling, Lunn, & Iacozza, 1999:302), as confirmed by (Amstrup et al., 2006:slide 44; Rosing-Asvid, 2006).

6 **Moreover, nowhere in the news coverage:** (IUCN Species Survival Commission, 2001:22).

6 **In 2006, a polar-bear:** (Taylor, 2006).

6 **Yes, it is likely that disappearing:** The Arctic Climate Impact Assessment finds it likely that disappearing ice will make polar bears take up "a terrestrial summer lifestyle similar to that of brown bears, from which they evolved." It talks about the "threat" that polar bears would become hybridized with brown and grizzly bears (Berner et al., 2005:509).

7 **In general, the Arctic Climate Impact:** "While there will be some losses in many arctic areas, movement of species into the Arctic is likely to cause the overall number of species and their productivity to increase, thus overall biodiversity measured as species richness is likely to increase along with major changes at the ecosystem level" (Berner et al., 2005:997).

7 **It will have less polar desert and:** (Berner et al., 2005:998).

7 **The assessment actually finds that:** (Berner et al., 2005:256).

7 **We are being told that the plight:** (Eilperin, 2004).

7 **We will later see that realistically:** This is based on a simple model starting in 2000 with a population of one thousand, a reduction of 1.5 percent (fifteen bears first year), and the full Kyoto Protocol reducing global warming by about 7 percent in 2100 (Wigley, 1998).

CHAPTER 2

10 **So let's start by asking the crucial:** (Karl & Trenbergh, 1999; Mahlman, 1997).

10 **Several types of gases can reflect or:** In this book, we will primarily

discuss CO_2, since it makes up 60 percent of the present extra heat-trapping gases and is expected to constitute an even larger part in the future. In 2100, CO_2 is expected to constitute anywhere from 68 percent of total forcing in scenario A2 to 97 percent in scenario B1 (IPCC, 2001a:403).

10 **The basic greenhouse effect is good:** (IPCC, 2001a:89).

11 **As natural processes only slowly remove:** (IPCC, 2007b:2.3.1).

11 **Whereas the developing world now is responsible:** Developing countries emitted 10.171Gt of the global 26Gt in 2004 (IEA, 2006b:513, 493). OECD countries were responsible for 51 percent in 2003 (OECD, 2006:148). Wigley estimates 29 percent from industrialized countries (OECD, 1998:2286). IPCC emissions scenarios range from 23 percent in the business-as-usual A1 scenario to 36 percent (Nakicenovic & IPCC WG III, 2000).

11 **In its "standard" future scenario:** (IPCC, 2007a:14; IPCC, 2007b: fig. 10.3.1) of A1B, described as the business-as-usual scenario (Dai et al., 2001).

12 **Globally, winter temperatures have increased much:** (Alexander et al., 2006; Easterling et al., 2000; Vose, Easterling, & Gleason, 2005).

12 **Moreover, winter temperatures have been:** (Michaels et al., 2000).

12 **Not surprisingly, this has:** (Easterling et al., 2000:419).

12 **However, with most warming heating up:** (Plummer et al., 1999).

12 **For the United States, the maximum temperatures:** (Easterling et al., 2000:419).

12 **In the Central England Temperature series:** (Horton, Folland, & Parker, 2001; Jones et al., 1999).

13 **Whenever there is a heat wave:** (Edwards, 2006; Vergano, 2006).

13 **As one environmentalist:** Climate scientist Bill Chameides of the conservation group Environmental Defense (Vergano, 2006).

13 **Famously, the chief scientific advisor:** (Lean, 2004). This line of thinking is also expressed by William Collins of the National Center for Atmospheric Research when he says: "Scientists are sure that we're changing the climate for the foreseeable future. What we're not sure about is whether or not we'll be able to live with those changes" (Edwards, 2006).

13 **In Al Gore's words:** (Gore & Melcher Media, 2006:75).

13 **The IPCC finds:** (IPCC, 2007b:chap. 10 summary).

13 **In the wintertime, temperatures:** (IPCC, 2007b:fig. 10.3.6).

13 **We will see a marked decrease in:** (IPCC, 2007b:chap. 10 summary).

14 **By the end of the century, we:** The model referred to in (IPCC, 2007b:10.3.6.2; Weisheimer & Palmer, 2005). They write two to three years and estimate their 5 percent extreme to increase to "around

40% or so," but the straight average of their numbers in table 2 gives 34 percent and 37 percent, around 1:3 years.

14 **In areas where there is one cold:** A 76 percent reduction in 2050 (Vavrus et al., 2006), referred to in (IPCC, 2007b:10.3.6.2).

14 **In the U.S. 2005 Climate:** (Ebi et al., 2006). See also (Basu & Samet, 2002; McMichael, Woodruff, & Hales, 2006), which talk *only* about heat-related deaths.

14 **For almost every location:** (Martens, 1998).

14 **However,** what **the optimal temperature:** Based on the summary of the biggest European heat and cold study (Keatinge et al., 2000:672).

14 **If you live in Helsinki:** The last forty-four days lie within the 3°C optimum-temperature zone (Keatinge et. al., 2000:672).

15 **Again, the death toll from excess:** With a population of 3.1 million.

15 **With more than 3,500 dead:** (Vandentorren et al., 2004:1519).

15 **Another 7,000 died in Germany:** (Larsen, 2003).

16 **The green group:** (Larsen, 2003).

16 **Fueled by such:** (Chase et al., 2007). Of course, global warming has been happening long before 1979, but we have satellite coverage only since then, and almost all of the second half of the twentieth century's warming has taken place since then.

16 **However, the BBC recently ran:** (BBC Anon., 2006b).

17 **In Europe as a whole:** 207,000, based on a simple average of the available cold and heat deaths per million, cautiously excluding London and using WHO's estimate for Europe's population of 878 million (WHO, 2004a:121).

17 **However, about 1.5 million Europeans:** 1.48 million, estimated in the same way as total heat deaths.

17 **our data suggest that:** (Keatinge et al., 2000:672). "Short term" refers to the fact that after a temperature increase both heat and cold deaths will tend to revert to their previous values, because of acclimatization, and will probably do so within a generation or two.

17 **For Britain:** (Keatinge & Donaldson, 2004:1096; Langford & Bentham, 1995) likewise estimate nine thousand fewer cold deaths.

17 **Indeed, a paper trying:** (Martens, 1998:342).

18 **Several recent studies have looked at adaptation:** (Davis et al., 2003; Davis et al., 2002).

18 **The effect creates:** (Anon., 2006; Arnfield, 2003).

18 **The British meteorologist Luke Howard:** Tel Aviv (Saaroni et al., 2000), Baltimore and Phoenix (Brazel et al., 2000), Guadalajara (Tereshchenko & Filonov, 2001), Barrow (Hinkel et al., 2003), Shanghai (Chen et al., 2003), Seoul (Chung, Choi, & Yun, 2004), Milan

(Maugeri et al., 2002), Vienna (Bohm, 1998), and Stockholm (Moberg & Bergstrom, 1997).

19 **In downtown Los Angeles:** (Akbari, Pomerantz, & Taha, 2001).

19 **New York has a similar nighttime:** (Rosenzweig et al., 2006).

19 **Over a short twelve years, the:** (Streutker, 2003).

19 **While the daytime temperature of:** (Hung et al., 2006:41).

20 **First, many of these:** We have to remember that the 2.6°C rise is a global average, and it is likely to be greater on land; moreover, many of the urban-heat-island figures are valid only for minimum or maximum temperatures. Yet for a city like Bangkok, the mean day temperature is 6°C higher throughout the dry season, and Tokyo ranges between 3 and 12°C higher throughout the year—thus still at least on scale with the 2.6°C average increase. (Khandekar, Murty, & Chittibabu, 2005:1573) note that Tokyo temperature over the past 150 years has increased 4°C.

20 **It is likely that for many cities:** For example: "On local and regional scales, changes in land cover can sometimes exacerbate the effect of greenhouse-gas-induced warming, or *even exert the largest impact on climatic conditions.* For example, urban 'heat islands' result from lowered evaporative cooling, increased heat storage and sensible heat flux caused by the lowered vegetation cover, increased impervious cover and complex surfaces of the cityscape" (emphasis added; Patz et al., 2005:310).

20 **This also does *not* deny:** For London, the urban-heat-island effect is estimated to add 0.26°C and fifteen days of intense urban heat per year by 2080 (Wilby, 2004:5).

20 **Thus, the sun's energy goes into:** (Greater London Authority, 2006).

21 **For London, that could decrease temperatures:** (Greater London Authority, 2006).

21 **Although it may seem almost comically:** (Synnefa, Santamouris, & Livada, 2006).

21 **In London, this could lower heat:** (Greater London Authority, 2006).

21 **Such a program:** The authors don't calculate the total cost, but given trees at $45, reroofing when necessary at $25 per hundred square meters, and painting of asphalt at $29 per hundred square meters, we get a total of $1.17 billion (Rosenfeld et al., 1998).

22 **At the moment, the only:** See (Grubb, 2004); for text, see (UNFCCC, 1997). A British government minister says: "The government's view is that Kyoto is the only game in town" (Dalyell, 2004).

22 **It has been championed by:** (Blakely, 1998); see also (Gore & Melcher Media, 2006:282–83, 288–89). Rob Gelbspan sees it as one of the few

causes for optimism (Gelbspan, 2004:x). For (Leggett, 2001), it constitutes the culmination of his epic story.

22 **Here it was decided that the industrial:** (Wigley, 1998:2286). The industrial nations are the so-called Annex I countries.

22 **The temperature by 2050:** Based on (Wigley, 1998) and a 4.7°F increase over the century.

22 **This is why** *The Washington Post*: (Anon., 2004).

22 **Even its staunchest backers admit Kyoto:** (Milliken, 2004). "Our efforts to stabilise the climate will need, over time, to become far more ambitious than the Kyoto Protocol" (Blair, 2004a).

23 **In total, this means that Kyoto:** (Bohringer & Vogt, 2003:478) estimates that Kyoto allows 4 percent *higher* emissions than what is expected—a bit like driving a car where the speed limit is slightly higher than the car's maximum speed.

23 **Realistically, the effective outcome will be:** A 0.7 percent reduction per (Bohringer & Vogt, 2003:481), about 0.8 to 1 percent per (Nordhaus, 2001:1283; Nordhaus, 2006c:fig. 4).

23 **If no other treaty replaces:** Based on the simplified assumptions of a 1 percent reduction in emissions over five years (2008–12) compared to the base run, modeled on the DICE-99 model (Nordhaus & Boyer, 2000).

23 **Lu Xuedu, deputy director of:** (Graham-Harrison, 2006); see also (Zhang, 2000).

24 **In estimating the cost of Kyoto:** Assuming smart trading (Weyant & Hill, 1999). Also quoted in (IPCC, 2001c:537; Golub, Markandya, & Marcellino, 2006:522).

24 **Depending on Russia and the future of:** Five billion to ten billion dollars (Dagoumas, Papagiannis, & Dokopoulos, 2006:37) is somewhere between 0.2 and 0.4 percent of GDP, due to expectations of future commitments and depending on the ability of Russia to limit the supply of permits, making them more expensive (Manne & Richels, 2004:453). If the EU were to make the cuts itself, this would still be very expensive, somewhere between 0 and 5 percent of national GDPs (Viguier, Babiker, & Reilly, 2003:479).

24 **Oil by itself is the:** (Craig, Vaughan, & Skinner, 1996:135).

24 **At more than $1.5 trillion:** Estimated with 80.1 million barrels per day at $54.57 (average 2005) (EIA, 2006c:87), against a global GDP of $47.767 billion (IMF, 2006:189).

24 **Consequently, there is big money to be:** In the modeling literature, this is known as the autonomous energy efficiency improvement (AEEI) factor, often estimated at between 0.7 and 1 percent (Weyant, 1996:1007), which has been criticized by (Grubb, Kohler, & Ander-

son, 2002) but shown to be possible to incorporate in a more general model in (Schwoon & Tol, 2006).

24 **Whereas the United States in 1800:** The equivalent of a 0.77 percent annual efficiency improvement (Lomborg, 2001:126).

24 **The average car driven in America has improved its:** From 13.6 mpg in 1973 to 22.4 mpg in 2004 (EIA, 2006d:17).

25 **Likewise, home heating in Europe and:** Europe used 24 percent less energy per square meter in 1992 than in 1973; the United States used 43 percent less (Schipper, Haas, & Sheinbaum, 1996:184).

25 **Many appliances have become much more efficient:** (Schipper, Haas, & Sheinbaum, 1996:187; Jaffe, Newell, & Stavins, 1999:13). In Denmark electric home appliances have become 20 to 45 percent more effective over the last ten years (NERI, 1998:238).

25 **We heat each room more efficiently:** (Lomborg, 2001:79).

25 **Ingenuity still works, and people constantly find:** (Weyant, 1996). For EU households, despite energy-efficiency gains, consumption *increased* by 2 percent annually in the last decade (Almeida et al., 2006).

26 **In a recent global-warming awareness week:** (BA, 2006).

26 **Isn't this costless or even advantageous:** Economists hold that there is no such thing as a free lunch—costs are bound to occur somewhere along the line. That there should be profitable CO_2 emissions reductions thus means that not only are there free lunches, but in some restaurants you can get paid to eat! (Lomborg, 2001:312–13).

26 **Why, they ask, would you:** (AP, 2006b).

26 **The ministry's "Warm Biz":** (AP, 2006b).

26 **An academic study found:** (Metcalf & Hassertt, 1997).

26 **At a cost of one thousand:** Quoted in (Monbiot, 2006:xvi).

27 **Likewise, it is often:** (Beinecke, 2005; Hawkins, 2001; Sierra Club, 2007).

28 **One way to think about:** (Mendelsohn, 2004).

28 **How much harm will that ton do:** Estimating that each mom drives eight kilometers at the average rate of 0.916 pounds per mile, or 0.258 kilograms CO_2 per kilometer (EPA, 2000), converted at 3.7 kilograms CO_2 per ton of carbon (IPCC, Houghton, et al., 1990:364).

28 **One hundred twenty-five people are going:** The U.S. average conversion rate is 1,630 kilowatts per ton of CO_2 (EIA, 1999:2; EIA, 2002:4); the EU (marginal) conversion rate is higher, at about 2,000 kilowatts per ton of CO_2, as estimated in (Almeida et al., 2006), with a standby phone-charger average of 1.5 watts per hour (Almeida et al., 2006: fig. 14). Attenborough was promised about four times that amount, but with no references (BA, 2006).

28 **Three people will have hot:** (Postman, 2006) estimates 342 pounds of CO_2 for a two-minute hot shower.

28 **If we place a tax:** (EIA, 2006f) estimates 8.87 kilograms of CO_2 per gallon of motor gasoline, or 0.887¢ per gallon.

29 **In a global macroeconomic model:** This is about $390 million per year.

29 **So we might want to think twice:** Notice that the cost function is strongly nonlinear, since deeper carbon cuts get much more expensive.

30 **In the biggest review article:** (Tol, 2005).

30 **Second, he finds that with reasonable:** Fifty dollars per ton of carbon (Tol, 2005:2071).

31 **When I specifically asked him for his:** From the Environmental Assessment Institute we asked him in July 2005: "Would you still stick by the conclusion that $15/tC [tons of carbon] seems justified or would you rather only present an upper limit of the estimate?" He answered: "I'd prefer not to present a central estimate, but if you put a gun to my head I would say $7/tC, the median estimate with a 3% pure rate of time preference" ($7/tC = $1.9/tCO_2$). This is comparable to Pearce's estimate of $1–2.5/tCO_2$ ($4–9/tC), Pearce (2003:369).

31 **If we tax it at $85:** (Stern, 2006:287); likewise the EU considers 20 euros (about 25 dollars) per ton of CO_2 "affordable" (EU, 2001).

31 **This is no trivial loss—all:** (Nordhaus, 2006b), with a permanent $85 per ton of CO_2 ($314.50/tC), at discounted present-day (2005) value.

31 **This is more than three times the:** $11.678 trillion (OECD, 2005:13).

31 **For the United Kingdom, the marginal cost:** (Pearce, 2003:377–78) estimates £45/tC, at 1£ = $1.93158 (25-11-06); this is equivalent to $23.70 per ton of CO_2.

32 **The models that estimate the:** (Nordhaus, 1992; Nordhaus, 1994; Nordhaus, 2001; Nordhaus, 2006a; Nordhaus & Boyer, 2000; Nordhaus & Yang, 1996; IPCC, Bruce, et al., 1996:385).

32 **These integrated models:** (IPCC, Bruce, et al., 1996:189; Nordhaus & Boyer, 2000:4–35).

33 **Costs are expressed:** (IPCC, Bruce, et al., 1996:187).

33 **For the full Kyoto Protocol:** (Nordhaus, 2006d). Compare with (Nordhaus, 2006c:25). Conceptually, this is the $150 billion annually, over the century, discounted to present-day (2005) value.

33 **There is an environmental benefit:** (Nordhaus, 2006d).

34 **Of course, when the time:** (Nordhaus, 2006c:10).

34 **With the United States out of:** (Nordhaus, 2006d).

34 **Thus, should the rich countries not:** (Nordhaus, 2006d).

34 **In the United Kingdom, the Kyoto CO$_2$:** (Nordhaus, 2006d).

35 **Currently, that cost is estimated at:** At £310/tC (Pearce, 2003:380).

36 **This is evident when we:** (Bohringer & Loschel, 2005) finds that most experts expect some reduction after 2012, but most expect fairly little.

36 **This is essentially the stated preference of:** This is equivalent to 1.5°C. The EU stated in a council decision in 1996 that temperatures should rise no more than 2°C above preindustrial levels—that is, no more than 1.2°C from today. (EU, 1996b), quoted in (EU, 2005:3). Of course, in 1991 the EU expected emissions to stabilize at 1990 levels by 2000 (EU, 1996a:iv).

36 **For every dollar spent, it will:** A benefit of a little less than $15 trillion in damages avoided, obtained at a cost of $84 trillion. See also the discussion in (Tol, 2007).

36 **This initiative sets a global carbon tax:** (Nordhaus & Boyer, 2000:7.6).

36 **It starts off with a carbon tax:** (Nordhaus, 2006d).

36 **Uniquely, it costs about $600 billion:** Notice that (Nordhaus & Boyer, 2000:7.7) use this scenario mainly as a benchmark: "This is not presented in the belief that an environmental pope will suddenly appear to provide infallible canons of policy that will be scrupulously followed by all. Rather, the optimal policy is provided as a benchmark for policies to determine how efficient or inefficient alternative approaches may be."

37 **A central conclusion from a meeting of:** (Nordhaus, 1998:18).

37 **In a review from 2006, the:** (Stern, 2006:298). See p. 135 for a critique of Stern's update.

37 **Thus, as one academic:** (Kavuncu & Knabb, 2005:369, 383).

38 **The first complete survey for:** (Bosello, Roson, & Tol, 2006:582).

38 **In total:** Cardiovascular death is more pronounced by cold than heat, thus we clearly see statistically more cardiovascular death with cold. The mechanisms are still not clearly delineated but several are seen as plausible (Cagle & Hubbard, 2005). With colder weather, the body restricts blood flow to the periphery to conserve heat, thereby leading to higher heart stress, decreasing thresholds for angina, and increasing the risk of dislodging a vulnerable plaque, which could lead to thrombosis. Cold also leads to more concentrated blood, causing more heart stress and increased blood pressure. Heat, on the other hand, might cause air (particulates and ozone) to aggravate respiratory mortality (O'Neill, Hajat, Zamobetti, Ramirez-Agular, & Schwartz, 2005).

39 **in the central estimates of the model:** (Tol, 2002b:154–55). Notice, though, that the difference between avoided cold deaths and heat deaths will gradually diminish toward 2200.

40 **Roughly that means paying almost one hundred dollars:** Thirty life-years at an average cost of sixty-two dollars (Hahn, 1996:236).

41 **We are being told by respected scientist:** (Lovelock, 2006a; Lovelock, 2006b). While Lovelock is technically correct—billions of us will die, since only very few of the six billion alive today will live to 2100—clearly his statement is that most of us will die much before our time. In (Lovell, 2006) he specifies that "a hot earth couldn't support much over 500 million."

41 **This is far beyond the pale of:** Yet Lovelock is being applauded by people such as Sir Crispin Tickell and Al Gore. Sir Crispin talks about how Lovelock gives "a marvellous introduction to the science" (Lovelock, 2006b:xvii), and Gore says, "Lovelock is truly a visionary"(Dana, 2006), clearly not challenging his science, though Gore believes that the political system will prove better than Lovelock fears.

41 **Second, when we talk:** (Helm, 2003) is an overview of climate-change policy, yet it has *no* mention of anything but carbon reductions and the political implications of this.

41 **We ought at least to consider adaptive:** (Goklany, 2006:314).

42 **Take a look at the World Health:** (WHO, 2002:224; Lopez et al., 2006). WHO's estimate is based on temperatures being 0.65°F higher today than in 1970.

42 **Thus, policies reducing the total problem:** (Goklany, 2006:322).

42 **We asked some:** (Lomborg, 2004; Lomborg, 2006). You can see more at http://www.copenhagenconsensus.com.

43 **But each expert didn't just state that:** In principle, both cost and benefit are measured equally into the future; in reality, the cost is often incurred over the next few years (as would any solution to cut CO_2 now or distribute mosquito nets now), but the benefit is measured over a much longer period yet still determined by the length of the models.

43 **Some of the top priorities also:** But notice that this need not have been so, because the Nobelists made the priority according to where we could do the most *extra* good for *extra* funds. Thus, there could have been small, obscure illnesses where the payoff would have been even better.

45 **For $27 billion, we can:** (Lomborg, 2004:104).

45 **Investing $12 billion:** (Lomborg, 2004:404–5).

45 **Yet an investment of $13:** (Lomborg, 2004:109; Lomborg, 2006: 26–27).

46 **They placed malnutrition:** (Lomborg, 2004:647)

46 **They came out with:** (Copenhagen Consensus, 2006). Kyoto is 23, and the other proposals ranked 37 to 40 out of 40.

47 **When we look into the:** MESSAGE A1 (Nakicenovic & IPCC WG III, 2000).

48 **Even the very worst-case scenario envisions:** This is the A2 scenario, with very high population growth and low economic growth.

48 **In this very unlikely case:** (Maddison, 2006).

49 **In one recent survey in Australia:** (Roy Morgan Research, 2006).

49 **In another survey, the United States:** (Chicago Council, 2006a:14, 16). Only India found "combating world hunger" slightly more important than "improving the global environment."

49 **South Korea put it first on:** (Chicago Council, 2006b:68).

49 **At stake is the survival of:** (Gore & Melcher Media, 2006:13).

50 **The climate crisis:** (Gore & Melcher Media, 2006:13). See also Gore: "I believe there is a hunger in the country to be part of a larger vision that changes the way we relate to the environment and the economy" (Dana, 2006).

50 **He explains how global:** (Gore & Melcher Media, 2006:13, 291).

51 **To be fair, Gore:** (Gore & Melcher Media, 2006:13).

51 **We have to imagine them asking:** (Gore & Melcher Media, 2006:13).

CHAPTER 3

53 **Over the past millennium:** Here based on the most inclusive temperature reconstruction of (Moberg et al., 2005),

53 **Between about 900 and 1200:** (Hughes & Diaz, 1994).

53 **The warmer climates and:** (Dillin, 2000), followed by (EB, 2006c).

54 **In Alaska, the mean temperature was:** (EB, 2006c).

54 **Evidence from a wide range of:** (Matthews & Briffa, 2005; Reiter, 2000).

54 **The Arctic pack ice extended so far:** (Reiter, 2000).

54 **Many European springs and summers were outstandingly:** (Burroughs, 1997:109).

54 **The harsh winters were captured by the:** (Reiter, 2000).

54 **Possibly the worst winter in France, in:** (Le Roy Ladurie, 1972:68), with population around 1700 estimated at twenty-one million (Le Roy Ladurie, 1972:68).

54 **Likewise in China, warm-weather crops:** (Reiter, 2000).

55 **Many have seized on pictures of these:** (Gore & Melcher Media, 2006:42–59; Pearce, 2005b).

55 **In Switzerland, there have been twelve:** (Joerin, Stocker, & Schluchter, 2006).

55 **One of the best-studied glaciers in:** While not entirely disappearing more than twice, they have names from Bjornbreen I to VI (Matthews et al., 2005).

55 **In fact, most glaciers in the:** (IPCC, 2007b:box 6.3).

55 **While glaciers since the last ice age:** (Oerlemans, 2000).

55 **It is estimated that glaciers around 1750:** (EB, 2006d).

55 **When Bjornbreen peaked around 1800, it:** (Matthews et al., 2005: 31); notice that this is only a schematic.

55 **The best-documented overview of glaciers shows:** (Oerlemans, 2005).

55 **The perfect glacier icon:** (Kaser et al., 2004).

55 **When Ernest Hemingway published:** (Kaser et al., 2004:330).

56 **the central theme from the:** (Kaser et al., 2004:331).

56 **Furthermore, Kilimanjaro has not lost:** (Kaser et al., 2004).

56 **In the latest satellite study, it:** (Cullen et al., 2006).

56 **In a video with:** (Greenpeace, 2001). *Rolling Stone* even displayed— without a hint of irony—a picture of Kilimanjaro as its first image of the "planetwide damage caused by global warming" and subtitled the picture from 1970 "Before the Warming: Kilimanjaro, 1970" (*Rolling Stone*, 2007).

56 **This is the price we pay:** (Reuters, 2001).

56 **When Greenpeace informs us:** (Thijssen, 2001).

57 **But what I do know:** (Ijumba, Mosha, & Lindsay, 2002; Richey, 2003; Soini, 2005:316; Vavrus, 2002).

57 **They are the biggest ice mass outside:** (Barnett, Adam, & Lettenmaier, 2005:306; Gore & Melcher Media, 2006:58).

57 **A stable glacier is not a continuous:** (Coudrain, Francou, & Kundzewicz, 2005:930).

57 **In this way, melting glaciers provide:** (Barnett, Adam, & Lettenmaier, 2005).

57 **The concern is that if:** (Barnett, Adam, & Lettenmaier, 2005:307).

58 **First, with glacial melting, rivers:** Up to 28 percent more in summer (Singh, Arora, & Goel, 2006:1991–92). Notice that this is for glacier-fed rivers, whereas snow-fed rivers will see a decrease (Singh & Bengtsson, 2005); for a basin with both glacier and snow feed, "reduction in melt from lower zones is counterbalanced by the increase in melt from upper zones" (Singh & Bengtsson, 2004:2382).

58 **Glaciers in the Himalayas have been declining:** (Lehmkuhl & Owen, 2005; Ruhland et al., 2006). At the same time, there seems to have

been less Himalayan snow accumulation since 1840 due to the weakening of the trade winds over the Pacific (Zhao & Moore, 2006).

58 **But with continuous melting, eventually the:** The simulations the IPCC refers to do not see a complete reduction of glaciers by midcentury but rather a 60 percent reduction (IPCC, 2007c:3.4.1; Schneeberger et al., 2003).

58 **Certainly we did so:** By 1700, France's forests had been reduced in size by more than 70 percent compared to 1000 C.E. The United States cut down about 30 percent of its original forest area, mostly in the nineteenth century (UNECE, 1996:19, 59).

58 **Since 1961 in the Karakoram and Hindu Kush:** (Fowler & Archer, 2006).

59 **The consequence has been a reassuring thickening:** (Fowler & Archer, 2006:4291).

60 **This worry is perhaps not surprising, since:** (EB, 2006a).

60 **Many commentators powerfully exploit:** (McKibben, 2004).

60 **When sea levels rise:** (Parkinson, 2006:42).

60 **Over the past forty years, glaciers:** (IPCC, 2007b:table 5.5.2). To most people, expansion of water seems unlikely to raise sea levels. Economist Richard Tol frequently uses the example of a cup of coffee, which does not shrink perceptibly when it cools; then he reminds people that the ocean is deep and that a 0.1 percent expansion of a cubic kilometer of water yields one meter of sea-level rise.

60 **In its 2007 report:** (IPCC, 2007b:10.6.5). Notice that (IPCC, 2007a) shows a midpoint of 38.5 centimeters.

60 **Since 1860, we have experienced a sea-level:** 11.4 inches since 1860 (Jevrejeva et al., 2006).

60 **it is also important to realize that:** The 1996 projection was 38 to 55 centimeters (IPCC & Houghton, 1996:364); 1992 and 1983 EPA estimates are from (Yohe & Neumann, 1997:243, 250).

61 **A cover story of *U.S. News:*** (Shute et al., 2001).

61 **Yet sea-level increase:** (Matthews, 2000).

61 **Moreover, with sea-level changes occurring:** (Yohe & Neumann, 1997).

61 **The IPCC cites the total cost for:** (IPCC, 2001b:396).

61 **Considering that the adequate protection costs for:** (EDD, 2006a: 11; EDD, 2006b).

62 **In a very moving section:** (Gore & Melcher Media, 2006:196–209).

62 **Well, technically, Al Gore is:** (Gore & Melcher Media, 2006:196).

62 **He is simply positing a hypothetical and:** Yet he also says: "First of all, this is not the worst case. The worst case, you don't want to hear!

I think I'm right down the middle and in fact, the scientific community has validated the science in this film, and, for example, the six metre, six to seven metre sea level rise—that would come if Greenland broke up and slipped into the sea. It would come if west Antarctica, the portion that's propped up against the tops of islands with the warmer sea coming underneath it, if it went. If both went, it would be 12 to 14 metres" (Denton, 2006).

62 **The UN estimates that over the century:** (IPCC, 2007b:fig. 10.6.1).

62 **Melting glaciers and ice caps will contribute:** (IPCC, 2007b: fig. 10.6.3). It is actually 8.8 centimeters, but the 0.8 seems to get lost somewhere in the sums.

63 **Likewise, Greenland is expected to contribute:** (IPCC, 2007b: fig. 10.6.4).

63 **However, as the world warms:** (IPCC, 2007b:fig. 10.6.4).

63 **So where are the nineteen:** (IPCC, 2007b:fig. 10.6.1).

63 **Likewise, they are not in the melting:** (IPCC, 2007b:table 4.1.1).

63 **If Antarctica entirely slipped into the ocean:** (IPCC, 2007b:table 4.1.1).

63 **Whereas the Antarctic is surrounded by ice:** (Huybrechts & de Wolde, 1999).

63 **Greenland is, as the IPCC:** (Johannessen et al., 2005; Zwally et al., 2005).

63 **Some analyses have shown more:** (Cazenave, 2006; Chen, Wilson, & Tapley, 2006; Howat, Joughin, & Scambos, 2007; Kerr, 2007; Luthcke et al., 2006; Murray, 2006; Velicogna & Wahr, 2006). (Shepherd & Wingham, 2007) estimates Greenland losing 100Gt per year or 0.28 millimeters per year.

63 **In a recent overview of all the:** (Oerlemans et al., 2005:235).

64 **In another overview, all:** (Gregory & Huybrechts, 2006:1721).

64 **The IPCC estimates that the very:** (IPCC, 2007b:fig. 10.6.4.3) has 0.2 meters, using (Parizek & Alley, 2004:1024), which gives only twenty-one centimeters at eight times the preindustrial CO_2 level— or two to four times higher than any of the IPCC scenarios by 2100. See also (Gregory & Huybrechts, 2006:1727).

64 **In 2006, we finally got:** (Vinther et al., 2006a; Vinther et al., 2006b).

64 **The temperature increases we see:** (Chylek, Dubey, & Lesins, 2006).

64 **The warmest year in Greenland was:** (Vinther et al., 2006b).

64 **Antarctica's gigantic ice sheet began:** (Zachos et al., 2001:688).

64 **The ice sheet is on average more:** (Parkinson, 2006:35).

64 **During the last ice age:** (Anderson et al., 2002; Bindschadler, 2006;

Huybrechts & de Wolde, 1999:2172) estimate Antartica contributing almost four centimeters per century to sea levels in stable mode.

65 **Here it is warming, while the:** (Chapman & Walsh, 2005; Humlum, n.d.; Monaghan & Bromwich, 2006).

65 **The South Pole has seen its temperature decline:** A decline of 0.4°C till 2006 by regression (GISS, 2006).

65 **However, the Antarctic:** (Marshall et al., 2006; Vaughan et al., 2003:266).

65 **In his film, Al Gore shows us:** (Gore & Melcher Media, 2006:182–83).

65 **The significance of this breakup:** Right after Gore's discussion of the Larsen B breakup, he shows us a large picture of high tides washing in over Tuvalu (Gore & Melcher Media, 2006:186–87).

65 **Studies show that in the:** (Pudsey et al., 2006).

65 **It is likely that the Larsen area:** (Pudsey et al., 2006:2375; Vaughan et al., 2001).

65 **Moreover, the breakup:** (Greenpeace, 2006b) points out it nonetheless is "a dramatic reminder of the effects of warming in the area."

65 **While it probably:** (Cook, Fox, Vaughan & Ferrigno, 2005; Parkinson, 2006).

66 **The precipitation on the Antarctic Peninsula:** (Turner et al., 2005) shows increasing precipitation. (Wingham et al., 2006:1629; Zwally et al., 2005:512) show large and increasing accumulation on the peninsula. (Morris & Mulvaney, 2004) shows that with conditions over the past thirty years 2°C temperature increases would mean about 0.012 millimeters per year less sea-level rise. They do, however, expect that sea-level increases from increased ablation could happen, and it would then be necessary to include increased precipitation explicitly.

66 **While most of Antarctica is too cold:** (Gregory & Huybrechts, 2006:1721).

66 **Al Gore also shows us:** (Gore & Melcher Media, 2006:178–79).

66 **Its population was:** (Barbraud & Weimerskirch, 2001:184). Gore claims a 70 percent decrease but provides no reference.

66 **This could possibly be linked to climate:** Only winter temperatures increased in the 1970s, which should actually have been a benefit for the penguins, because it would have made hatching more successful (Barbraud & Weimerskirch, 2001:185). Moreover, winter temperatures declined again in the 1980s and beyond, without population increases.

66 **However, it is but a single:** (Australian Government Antarctic Division, 2003).

66 **Some of the largest colonies contain more:** (Woehler & Croxall, 1997:44) notes Kooyman found that colonies in the Ross Sea may be increasing. (Kooyman, 1993) estimates both Cape Washington and Coulman Island as having around twenty thousand fledgling chicks (or breeding pairs). In 1964, Cape Washington was registered to have only about 2,500 to 3,800 pairs (Wilson, 1983:5).

66 **The IUCN estimates that there are almost:** (BirdLife International, 2004). (Grzimek, n.d.) even lists the population as stable or increasing.

66 **Moreover, the other main Antarctic penguin:** (Jenouvrier, Barbraud, & Weimerskirch, 2006) shows a 1.77 percent increase over the period.

67 **While studies are uncertain as to whether Antarctica:** (IPCC, 2007b:fig. 4.6.2.2; Gregory & Huybrechts, 2006:1721).

67 **So what will be the:** For example, in the 2001 IPCC report; see (Lomborg, 2001:289–90). This is also true in the new policy summary of IPCC WGII, which states all its main findings in its only table, bluntly admitting: "Adaptation to climate change is not included in these estimations" (IPCC, 2007d).

67 **Let us look at some:** (Nicholls, 2004; Nicholls & Tol, 2006), using the A1FI scenario (which actually makes the argument stronger than A1B).

67 **This is evident in the United States:** (Pielke & Landsea, 1998: fig. 3)

67 **A more clear-cut example is:** (Waltham, 2002:95).

69 **However, such a future would:** This is based on the two standard IPCC scenarios, A1 and B1 (BEA, 2006a; Nicholls, 2004:72) in 2005 dollars. Essentially the A1 expects a world focused on economic growth and consequently higher growth rates, whereas the B1 sees a world more focused on environmental solutions with less CO_2 emissions but also less economic growth.

69 **Despite one-third less sea-level rise:** No upgrade for B1 (about two million flooded) and an upgrade for A1 (less than one million flooded) (Nicholls & Tol, 2006:1084).

69 **For more than 180:** (Nicholls & Tol, 2006:1088), estimated for 2085. Notice that low-lying undeveloped coasts in places such as Arctic Russia, Canada, and Alaska are expected to be undefended. Notice that the numbers presented are for loss of dry land, whereas up to 18 percent of global wetlands will be lost.

69 **The most affected nation will be Micronesia:** Micronesia (CIA, 2006).

69 **If nothing is done, Micronesia will:** (Tol, 2004:5).

70 **The 77 percent land:** "A cursory comparison of protection costs and the costs of dryland loss indicates why protection levels are so high" (Tol, 2004).

71 Tony Blair tells us that "sea: (Blair, 2004a).

71 If everyone, including the United States and: (Wigley, 1998).

71 Likewise, Greenpeace tells us that: (Greenpeace, 2006b).

72 The National Resources Defense Council tells us that: (NRDC, 2006).

72 Extreme weather events are predicted to become: (FOE, 2006).

72 Greenpeace tells us that "there is strong: (Greenpeace, 2006a). (Greenpeace, 2004) tells us that "as climate change gathers pace, devastation caused by extreme weather is becoming more common."

73 Robert F. Kennedy Jr.: (Kennedy, 2005).

73 Just a day after Katrina wrecked Louisiana: (Gelbspan, 2005).

73 So has global warming caused: (WMO-IWTC, 2006a; WMO-IWTC, 2006b; WMO, 2006). This was concluded in December 2006, whereas the material deadline for IPCC was earlier in the year.

73 Though there is evidence both for and: (WMO-IWTC, 2006b).

73 Al Gore is incorrect when: (Gore & Melcher Media, 2006:92).

74 The recent increase in societal impact from: (WMO-IWTC, 2006b).

74 Gore ends his discussion on hurricanes by: (Gore & Melcher Media, 2006:202).

74 He tells us that by 2040: (Gore, 2006a).

74 The global costs of climate-related: Personal communication from Angelika Wirtz Geoscience Research Group, Munich Re.

74 There are two: $6.465 billion in 2005 versus $2.519 billion in 1950 (UNPD, 2006:5). Average income of $9,233 in 2005 versus $2,803 in 1950 (Worldwatch Institute, 2006:53).

75 each of us: (Pielke, 1999; Swiss Re, 1999:8).

75 In Florida, Dade and Broward counties: (Pielke & Landsea, 1998).

75 Hurricane Andrew in 1992, breaking all: (Pielke & Landsea, 1998; Pielke, 2006; Pielke et al., 2007).

75 Hitting just north of where Hurricane Andrew: (NOAA, 2006).

76 The same story goes for: (NOAA, 2006).

76 Had it hit today, it would: Notice that both the Great Miami Hurricane and the Galveston hurricane caused many deaths—the Galveston one being the deadliest in U.S. history, with eight thousand dead. Had these storms hit today, obviously better warning systems would have evacuated much of the city and perhaps forestalled many of the deaths, but the material damage would have been equally severe.

76 The Association of British Insurers found that: (Association of British Insurers, 2005). They make a number of other statements about the effectiveness of climate-change policies but do not compare them to the effects of socioeconomic factors.

76 **According to one current insurance-industry report:** (*Insurance Journal*, 2006).

76 **But how can we then:** Based on (Pielke, 2005; Pielke, Klein, & Sarewitz, 2000), this is an average of the three very similar climate increases and the A1 scenario social increase.

77 **If we managed to get everyone:** With Kyoto we would see 6 percent lower temperatures in 2050, leading to 0.6 percent less of an increase than the 10 percent envisioned in (Pielke, 2005; Pielke, Klein, & Sarewitz, 2000).

78 **This would abet evacuation plans, community:** The following examples are mainly from (Pielke, Klein, & Sarewitz, 2000).

78 **After Hurricane Katrina, one insurance company:** (Mills & Lecomte, 2006:16).

79 **Often, simple structural measures:** (Stern, 2006:420).

79 **What caused the tragedy in New Orleans:** (McCallum & Heming, 2006).

79 **Models and exercises had repeatedly shown that:** (Congleton, 2006; Travis, 2005).

79 **One climatologist pointed out:** (McCallum & Heming, 2006:2113).

80 **It is perhaps sobering to:** (Tol, 2002a:49).

80 **we know effective:** (Sarewitz & Pielke, 2005).

80 **Those who justify:** (Sarewitz & Pielke, 2005).

81 **Unusually severe floods:** (Mitchell, 2003).

81 **After the severe flood of Prague and:** For a description, see (Petrow et al., 2006). On the call for Kyoto: "Blair, Chirac and German Chancellor Gerhard Schröder urged final ratification of the Kyoto Protocol on climate change, recalling floods that hit central Europe last month" (Reuters, 2002).

81 **According to Schröder:** (Reuters, 2002; Xinhuanet, 2002).

81 **And yes, it is:** (Groisman et al., 2005; IPCC, 2007b:10.3.2.3, 10.3.6.1).

81 **Models also show that this will lead:** (Milly et al., 2002).

81 **There is also some evidence that increased:** (IPCC, 2007b:Q9.1). See also (Barnett et al., 2005) for general attribution of precipitation and (Bronstert, 2003; Huntington, 2006) for weak or no connection between floods and climate.

82 **This holds true in a global sample:** (Kundzewicz et al., 2005), contradicting the much smaller data set from (Milly et al., 2002).

82 **This also holds true for the smaller:** (Svensson, Kundzewicz, & Maurer, 2005).

82 **When studying U.S. rivers:** (Small, Islam, & Vogel, 2006) confirms (Lins & Slack, 1999; Lins & Slack, 2005; USGS, 2005), though (Groisman, Knight, & Karl, 2001) finds a signal also in high stream flow.

82 **Likewise, in Europe, a study:** (Mudelsee et al., 2003).

82 **This is well correlated with:** (Thorndycraft et al., 2006).

82 **With much snow and a late thaw:** (Demaree, 2006).

82 **These floods have decreased sharply in the:** (Demaree, 2006:895–96; Pfister, Weingartner, & Luterbacher, 2006).

82 **Likewise, an analysis of the river:** (Mudelsee et al., 2006).

82 **Along the river Vltava in the Czech:** (Yiou et al., 2006).

83 **In general, casualties:** (Mitchell, 2003).

83 **Flooding has been pervasive:** (Mitchell, 2003).

83 **What does set our period:** (Brazdil, Kundzewicz, & Benito, 2006; Mitchell, 2003).

83 **The congressional Office of Technology:** (Pielke, 1999:419ff.).

83 **This would be fine, if flood:** (Pinter, 2005).

84 **Moreover, levees themselves tend:** (GAO, 1995:37).

84 **Imagine a river one thousand yards across:** (Larson, 1994).

84 **Flows that were fully contained within the:** (Pinter & Heine, 2005).

84 **As with hurricanes, we:** (BEA, 2006a; BEA, 2006b; Downton, Miller, & Pielke, 2005a; Downton, Miller & Pielke, 2005b; Pielke & Downton, 2000). Exponential trend lines: $y = 0.4871\exp(0.0308(x-1928))$ and $y = 172.01\exp(-0.0046\,(x-1928))$.

85 **The only large-scale:** Assuming linearity in temperatures, with A1FI temperature in 2100 at 4.49°C, B1 at 1.98°C, and Kyoto reduction at 7 percent of A1FI temperatures in 2100 (IPCC, 2001a:824; Wigley, 1998).

85 **This would also entail no public subsidies:** (Pinter, 2005).

86 **The Foresight studies show that the:** (Evans, Ashley, Hall, Penning-Rowsell, Sayers, Thorne, et al., 2004:217–18),

86 **From Kyoto, at:** Using the GDP estimates for the world scenario and the relative efficiencies explained above (Evans, Ashley, Hall, Penning-Rowsell, Saul, Sayers, et al., 2004:225; Evans, Ashley, Hall, Penning-Rowsell, Sayers, Thorne, et al., 2004:217–18).

87 **The Gulf Stream and its:** Including the North Atlantic current (EB, 2006b; Seager, 2006). In the text I use "Gulf Stream" for all of these currents. I also use it as a short and popular understanding for the so-called thermohaline circulation, because it is primarily wind driven, as is the Gulf Stream (Wunsch, 2002). "Thus the Gulf Stream, and hence the wind, rather than being minor features of oceanic climate are best regarded as the primary elements" (Wunsch, 2006).

87 **One day, the ice dam broke:** This is the story Gore tells (Gore & Melcher Media, 2006:151), although he gets the timing wrong (Barber et al., 1999; Meissner & Clark, 2006).

88 **Of course, there is no glacial:** In his movie, Gore looks at the North

Atlantic map and coyly asks, "Is there another big chunk of ice near there?" Then, observing Greenland, he says, "Oh, yeah"(Gore, 2006b).

88 **A team of modelers looked:** (Jungclaus et al., 2006).

88 **Although they see a reduction in the:** (Jungclaus et al., 2006). Their results seem to hold till at least 2200.

88 **The original concern over a Gulf Stream:** (Pearce, 2006:185).

88 **Europe's climate could:** (Calvin, 1998:47).

89 **In 2004, *Fortune* magazine revealed "the:** (Stipp, 2004; Townsend & Harris, 2004).

89 **The study sketches a "plausible":** (Schwartz & Randall, 2003).

89 **It is not hard to:** The movie's website linked to news reports from February 2004 about "a secret report prepared by the Pentagon" that warned climate change would "lead to global catastrophe costing millions of lives." Academic papers have also used the *Day after Tomorrow* reference see, for example, (Hansen et al., 2004).

90 **In the event 8,200 years:** (Barber et al., 1999; Wiersma & Renssen, 2006:73).

90 **Model estimates show the same kind of:** (Stouffer et al., 2006; Wood, Vellinga, & Thorpe, 2003).

90 **For comparison, the average temperature difference:** Rough calculation from 4°C digitized global-temperature map in (NCEP, 2006), with Europe to 40°E at 6.94°C and with Siberia, defined from 60°E and north of 50°, with −5.86°C. A similar estimate for Siberia comes from (FAO, 2001:chap. 27), with western Siberia −4°C, south Siberia −0.5°C, the Siberian plateau −12°C, and central Siberia −13.5°C, or a simple average of −7.5°C.

90 **In fact, "the only way:** (Wunsch, 2004).

90 **Over the past forty-seven years, scientists:** (Bryden, Longworth, & Cunningham, 2005; Kerr, 2005).

90 *National Geographic* **told us " 'Mini Ice:** (Owen, 2005).

90 *New Scientist* **chose the news as one of:** (Pearce, 2005c).

90 **The *Sydney Morning Herald* even saw a way:** (Connor, 2005; Henderson, 2005; Smith, 2005).

91 **If the Gulf Stream were really slowing:** (Pearce, 2005a; Schiermeier, 2006:259).

91 **The story is appealing:** (Schiermeier, 2006:258).

91 **In *Science* the headline ran:** (Kerr, 2006).

91 **In *New Scientist* the headline was:** (Merali, 2006).

91 **Unfortunately, it seems no other major:** Based on a Google search for "RAPID array observations Birmingham," " 'Rapid Climate Change Conference' Birmingham," and "RAPID Bryden Birmingham."

91 **This is also why the IPCC:** (IPCC, 2007b:question 10.2).

92 **But if—more plausibly—our:** See, for example, (Link & Tol, 2004).

92 **Europe will still experience:** (IPCC, 2007b: question 10.2). The IPCC writes "radiative forcing" instead of "CO_2 warming" and "MOC" instead of "Gulf Stream."

92 **Malaria in Vermont:** (McMichael et al., 2003; WHO, WMO, & UNEP, 2003).

92 **Green organizations, political parties, and:** (Khaleque, 2006; Lib-Dem, 2006:6; Tindale, 2005).

93 **Not surprisingly, a headline like:** (Plumb, 2003).

93 **WHO produced a:** (Campbell-Lendrum, Corvalán, & Prüss-Ustün, 2003).

93 **There, the authors set out to estimate:** The analysis was redone in 2005 with almost exactly the same results (Patz et al., 2005).

93 **While the authors spent:** "Climate change attributable deaths were calculated as the change in proportion of temperature-attributable deaths (i.e., heat-attributable deaths plus cold-attributable deaths) for each climate scenario compared to the baseline climate" (Campbell-Lendrum, Corvalán, & Prüss-Ustün, 2003:142). They have only old and very limited surveys for these results—for Europe they use (Kunst, Looman, & Mackenbach, 1993), which covers only the Netherlands, though a total European survey is available in (Keatinge et al., 2000).

93 **when they aggregated the:** "Relative risks for 2000 have been estimated as described above, and applied to the disease burden estimates for that year, with the exception of the effects of extreme temperatures on cardiovascular disease, for the reasons described above" (Campbell-Lendrum, Corvalán, & Prüss-Ustün, 2003:152). However, there are no "reasons described above" anywhere. In the 2005 update, although they mention both cold and heat deaths and attempt an estimate at heat deaths, they equally leave them out of the total (Patz et al., 2005:312).

93 **If we make a rough:** (CRU, 2006) shows 0.361°C change from the 1961–90 average in 2000; (WHO, WMO, & UNEP, 2003:7) estimates 0.4°C. The estimate comes from a linear extrapolation from (Bosello, Roson, & Tol, 2006), which estimates increases of 1.03°C from today's temperature. Taking the proportional (0.35 = 0.361/1.03) cold and heat deaths gives the numbers here. It also gives an estimate of all other illnesses of 193,000, which compares fairly well with WHO's 150,000.

94 **We see the same claims:** Three hundred to five hundred million in (WHO & UNICEF, 2005:xvii), 515 million in (Snow et al., 2005), com-

pared to almost two billion febrile episodes resembling malaria each year (Breman, 2001).

94 **Former UN secretary-general:** (Annan, 2006).

94 **The cover story:** (Shute et al., 2001).

94 **As in most stories, there is:** (Snow & Omumbo, 2006:197).

94 **We can determine that there will be about:** (Martens et al., 1999; van Lieshout et al., 2004), behind the claims of (King, 2004).

94 **The same models find:** (Arnell et al., 2002:439) interpolated from 1990 and 2025, with world population at 6.6 billion (USCB, 2007).

95 **This is more than ten times the:** (Reiter et al., 2004).

95 **When I first started reading about malaria:** (CDC, 2006; Reiter, 2000; Swellengrebel, 1950).

95 **But there is abundant evidence—from:** (Kuhn et al., 2003; Reiter, 2000). See also map of spread in 1900 in (Hay et al., 2004).

95 **Up till the 1940s:** (Boyd, 1975; CDC, 2004; Reiter, 2000:9).

95 **Dr. William Currie:** (Thompson, 1969:199).

95 **The permanent secretary of the California:** (Thompson, 1969:199).

96 **Experts estimated that for the nineteenth century and:** (Madden, 1945:2).

96 **In 1920, almost 2 percent of:** (USCB, 1999:875; Mégroz, 1937:353).

96 **This is why CDC headquarters:** (CDC, 2004).

96 **A long list of factors caused the:** (CDC, 1999:106; Konradsen et al., 2004).

96 **Perhaps tellingly, a 1944 analysis:** (Brierly, 1944).

97 **A recent European study:** (Kuhn et al., 2003).

97 **Likewise for the United States, an:** (Longstreth, 1999).

97 **A belt runs through the middle of:** (Guerra, Snow, & Hay, 2006; Hay et al., 2004; Snow et al., 2005).

98 **Moreover, concerns from Western governments:** (Beard, 2006; Rosenberg, 2004; Schapira, 2006; Walker, 2000).

98 **These populations are also:** (Jamison et al., 2006:3; Snow & Omumbo, 2006:205).

98 **Over the past twenty years:** (Snow & Omumbo, 2006:208).

98 **At the same time, temperatures have:** (Epstein, 2000; Epstein et al., 1998; Patz et al., 2005).

98 **There has been a substantial literature trying:** (Hay, Cox, Rogers, Randolph, Stern, Shanks, et al., 2002a; Hay, Rogers, et al., 2002; Hay, Cox, Rogers, Randolph, Stern, 2002b; Pascual et al., 2006; Patz et al., 2002; Reiter et al., 2004; Shanks et al., 2002). Personally, I found this statement from advocates of a warming link to malaria very telling: "The absence of a historical climate signal allows no inference to be drawn about the impact of future climate change on malaria in the

region" (Patz et al., 2002). So data should not tell us something about the future?

98 **A recent World Bank review:** (Snow & Omumbo, 2006:208).

98 **There are new and effective combination treatments:** (Purcell, 2006; Shanks, 2006).

99 **These are the models that show an:** (Arnell et al., 2002; Martens et al., 1999; van Lieshout et al., 2004). Arnell finds 289.5 million as an average among unmitigated scenarios. We here use Arnell, since he is the only one to publish a population at risk without climate, but he stays within the same framework and range of outcomes as the other articles.

99 **Second, the analysis does not take:** (van Lieshout et al., 2004:91): "This assessment will describe potential populations at risk based on the *current* level of adaptation to malaria."

99 **Of course, when developing countries go:** (World Bank, 2006:289).

99 **Finally, the models also disregard that:** (van Lieshout et al., 2004:97). See also (Hay, Guerra, et al., 2005; Utzinger & Keiser, 2006:530; van Lieshout et al., 2004:96–97).

99 **This is also reflected:** (Rogers & Randolph, 2000).

99 **They project an increase:** (Arnell et al., 2002:439).

100 **Thus, even if we could entirely:** That is, 289.5 million / 9.1095 billion.

100 **More realistically, adoption of the Kyoto Protocol:** That is, 289.5m/9109.5m * 7% (Wigley, 1998:2287).

100 **As a model team tells us:** Assuming a 550 ppm (parts per million) stabilization (Arnell et al., 2002:440).

100 **Compare this to current expectations:** (Mills & Shillcutt, 2004: 84–85).

100 **In comparison, that to a simple and cheap:** Calculated from five hundred million actual annual malaria cases in 2000 and proportional from there with the data in (Arnell et al., 2002:439).

100 **This policy will do about four hundred times:** This is also evident in (Tol, Ebie, & Yohe, forthcoming).

100 **In Zambia, more than:** (WHO & UNICEF, 2003:20).

100 **Likewise, poor households across twenty-two:** (WHO & UNICEF, 2003:28).

100 **Rural children—who are often also:** (WHO & UNICEF, 2003:28, 35).

100 **Rich children in general get more:** (WHO & UNICEF, 2003:35).

101 **Average annual personal income:** All in 2005 dollars (BEA, 2006a; Department of Commerce, 1982:54; Department of Commerce, 2006:D-71).

101 **In sub-Saharan Africa:** (World Bank, 2006:289).

101 **Studies show that when countries get:** (Tol & Dowlatabadi, 2001).

102 **A recent Associated Press story spends 774:** (Hanley, 2006).

102 **WHO points out in very blunt:** (Snow et al., 1999:6.2).

102 **However, WHO finds that the real:** Similar to the findings in (Shanks et al., 2005).

102 **It is also the first of the:** (UNDESA, 2006:5).

102 **Stories of how global warming will:** (McCarthy, 2005; Pullella, 2005).

103 **Likewise, the proportion of malnourished has:** (FAO, 2006:8; FAO, 2007; Fischer et al., 2005:2080; Fischer, Shah, & Velthuizen, 2002: 112–13; Grigg, 1993:50; Nakicenovic & IPCC WG III, 2000; WFS, 1996:1:table 3).

103 **A few large-scale:** (Fischer et al., 2005; Fischer, Shah, & Velthuizen, 2002; Fischer et al., 2002; Parry, Rosenzweig, & Livermore, 2005; Parry et al., 2004; Rosenzweig & Parry, 1994). As (Fischer et al., 2005) is the only recent attempt that uses a variety of climate models, this will be the central one used here.

103 **All models envision:** For A1 and increase from 1,800 Mt to 3,900 Mt (Parry et al., 2004:64).

103 **Thus, we will be able to feed:** (Fischer et al., 2005:2080).

103 **For the most pessimistic models and the:** This is the HadCM3 model with a high climate sensitivity (Fischer et al., 2005:2071) and the A1FI, which has absolutely the highest CO_2 concentration (Fischer, Shah, & Velthuizen, 2002:109).

104 **A lower climate impact and the most:** A2 and NCAR (Fischer, Shah, & Velthuizen, 2002:109).

104 **To put these numbers in:** (FAO, 2006:16).

104 **In the most negative:** It is important to point out that we expect lower growth rates in the future, but that is because of lower demand, not because of inherent production limits, as the A2 scenario clearly shows, enabling total cereal production of 4800 Mt (Parry et al., 2004:64).

104 **The total agricultural GDP:** (Fischer, Shah, & Velthuizen, 2002:109).

104 **However, agriculture will constitute less than:** At most $3.6 trillion of $380 trillion (Fischer, Shah, & Velthuizen, 2002:108; Nakicenovic & IPCC WG III, 2000).

104 **For both places, however, CO_2:** In this chapter, it is assumed that the full CO_2 effects are captured. (Long et al., 2006) raised doubts about this, but (Tubiello et al., 2007) shows that FACE studies are in line with previous studies. (IPCC, 2007c:5.4.1.1) also concludes: "Our assessment is that main crop simulation models, such as CERES, Cropsys, EPIC, SoyGrow, and main pasture models CENTURY and EPIC, are in line with recent findings—in fact a bit lower—by assuming crop yield increases of about 8–17%."

104 **In worst-case scenarios:** (Parry et al., 2004:64).

105 **However, this does not mean that:** (Fischer, Shah, & Velthuizen, 2002:96). The production increase could have been even larger without global warming.

105 **Even without global warming, imports for:** LCDs in A1 (Fischer, Shah, & Velthuizen, 2002:98).

105 **Global warming causes this import to increase:** (Fischer et al., 2005:2079).

105 **In the most likely scenario it means:** (Fischer, Shah, & Velthuizen, 2002:112).

106 **However, how many hungry:** (Parry et al., 2004:62) and see especially the A2 and B2 versus A2s and B2s in (Fischer, Shah, & Velthuizen, 2002:100): "The significance of any climate-change impact on the number of undernourished depends entirely on the level of economic development assumed in the SRES scenarios" (112).

106 **Even the modelers:** (Parry, 2004).

107 **Another way to see this:** Hunger is only 26 percent about increased food availability but much more about women's education, status, and health (Sanchez et al., 2005:22–23).

107 **As pointed out by the modeling team:** Left out "wealthy societies..." (Fischer, Shah & Velthuizen, 2002:112–13).

107 **If we implemented Kyoto, this would:** Twenty-eight million times 7 percent (Wigley, 1998:2287).

107 **We could focus on simple measures like:** (Sanchez et al., 2005:189).

107 **The UN estimates that we could reduce:** (UN Millennium Project, 2005:252) estimates the total extra cost of meeting the Millennium Development Goals at 0.44 to 0.54 percent of OECD countries GDP from 2005 to 2015—approximately $165 billion (OECD, 2005:13). (Sanchez et al., 2005:18) estimates the total cost of hunger reduction to be 5 to 8 percent of the total Millennium Development Goals cost, which is $8.25 billion to $13.2 billion. That seems to be fairly well reflected in the cost structure in (UN Millennium Project, 2005:244). The 229 million are the *extra* people saved—that is the difference between the numbers of expected hungry in 2015 without the extra effort, 749 million, and with it, 520 million (UN Millennium Project, 2005:259).

107 **If we look at the effect over:** One-eighteenth of two million (111,111), phased in from 2050 to 2080 and constant till 2100, averaged over the entire century.

108 **Each time our investment in climate:** Estimating the 229 million phased in from 2005 to 2015 and constant thereafter (this is a *very*

conservative estimate, since the economic growth and human well-being involved in avoiding 229 million hungry people will probably quickly make these gains self-sustaining), meaning 206 million over the century. (206 million / 39,000 = 5,282).

108 **Rather, the more we understand about:** (Gore & Melcher Media, 2006:117).

108 **We need to get better at using:** See the analysis in (Lomborg, 2001:149–58).

108 **In its 2006 world:** (UNESCO, 2006:45). Also, "there is enough water for everyone. The problem we face today is largely one of governance" (3).

108 **There is a water crisis:** (World Water Council, 2000:xix).

109 **The typical way to measure:** Used by the World Bank and many others (Ashton, 2002; Revenga et al., 2000; Serageldin, 1995; Simonovic, 2002; UNEP, 2000).

109 **It essentially looks at how much water:** (Arnell, 2004:37).

109 **One of the largest models from the:** (Arnell, 2004).

110 **This is because a warmer:** This is for the A1B scenario (Nohara et al., 2006:1081).

110 **It is important to note:** (Arnell, 2004:50) stresses that increases in runoff will generally occur during high flow seasons, but with a multiple of models this is true for the Amazon, Ganges, and Mekong but not for the Amu Dar'ya, Columbia, Danube, Euphrates, Lena, Mackenzie, Nile, Ob, Syr Dar'ya, Volga, and Yenisey (Nohara et al., 2006:1085–86). (Arnell, 2004:50) also mentions that "the watersheds that apparently benefit from a reduction in water resources stress are in limited, but populous, parts of the world, and largely confined to east and southern Asia: areas that see an increase in stress are more widely distributed," as if area counts besides the number of humans for human welfare. To me, this argument seems indefensible.

111 **We could bring basic water and sanitation:** About $10 billion per year from 2007 to 2015, from a range of global studies (Toubkiss, 2006:7); compare this to about $100 billion over the period in (Rijsberman, 2004:521). Four billion dollars per year forever is the equivalent of $10 billion from 2007 to 2015 at a 5 percent discount rate.

111 **This would avert almost:** (Hutton & Haller, 2004:25).

111 **The total monetary value of this benefit:** (Hutton & Haller, 2004:32). Surprisingly, perhaps, the biggest gain comes from sanitation, not water.

111 **But hasn't global warming done:** (Gore & Melcher Media, 2006:119).

111 **It is absolutely correct that the Sahel:** (Dai et al., 2004; Giannini,

Saravanan, & Chang, 2003). Of course, all that says is that the climate system is causally connected—what we then need is to show that the Indian Ocean's heightened temperatures are caused by global warming.

111 **However, when two sets of researchers:** (Hoerling et al., 2006; Lau et al., 2006).

111 **Of nineteen models, only eight:** (Lau et al., 2006:8).

111 **One research team concluded:** (Hoerling et al., 2006).

CHAPTER 4

115 **At one-sixtieth the cost:** Dealing with malaria by affecting a change in global warming through the Kyoto Protocol will probably save on the order of 0.1 percent of annual malaria deaths, averaged over the century (289.5 million/9.1095 billion * 7 percent/2) (Arnell et al., 2002:439; Wigley, 1998:2287), or about one thousand lives each year, (at one million deaths [Teklehaimanot et al. & UN Millennium Project, 2005:1]). In comparison, a targeted approach could cut malaria deaths by 75 percent, or 750,000 per year, averaged over the century. (Notice that, because of growth in population and climate effects, the actual averages would probably be about 1,400 and 850,000.) Moreover, the cost of Kyoto would be $180 billion annually, compared to just $3 billion annually for a targeted malaria policy (Teklehaimanot et al. & UN Millennium Project, 2005:2; Weyant & Hill, 1999). Kyoto would therefore save 140,000 people at sixty times the cost, whereas a targeted malaria policy would save more than 85 million.

116 **But perhaps more important, besides the:** (Sarewitz & Pielke, 2007).

117 **In March 2007, the EU members:** (EU, 2007c:12). Notice that promising such a goal is not the same as reaching it. In the same EU document, the EU actually lauds the accomplishments of its Lisbon Strategy from 2000, "aimed at making the European Union the most competitive economy in the world" (EU, 2007b). A central target here is achieving spending of 3 percent of GDP on R&D. Yet a recent LSE assessment shows that the target "will not be achieved by 2010" (CEP, 2006). While the EU average for R&D in 2000 was 1.86 percent, the latest figures from 2005 have *declined* to 1.84 percent (for EU-27; for EU-15, it went from 1.92 to 1.91 percent) (EU, 2007a).

117 **This would mean a 25 percent cut:** (IEA, 2006b:507).

117 **The cost would be about ninety:** Estimated using (Nordhaus, 2006d).

117 **CO$_2$ emissions over the:** (Bohringer & Vogt, 2003:478; EIA, 2006a; EIA, 2006b; EIA, 2006e; IEA, 2006a:II.4; IEA, 2006b:493, 529; Marland, Andres, & Boden, 2006).

118 **Those countries that are on paths toward:** (Sarewitz & Pielke, 2007).

119 **At the Earth Summit in Rio de Janeiro in:** (UNFCCC, 1992:4.2a).

119 **Most experts also expect that:** Seventy-eight climate policy experts expect a 10 percent reduction (Bohringer & Loschel, 2005).

120 **Not surprisingly, R&D in:** See (IEA, 2007; WDI, 2007), using the average of individual country percentages. (The average of total investment against total GDP is similar but highly influenced by Japan, which spends almost half of all its R&D in energy efficiency.) Here I use the same list of countries as (Runci, 2005), though without the United States: Canada, Denmark, France, Germany, Italy, Japan, the Netherlands, Spain, Sweden, and the United Kingdom, representing about 95 percent of industrialized country R&D investments in 2003. Private R&D data for the United States are from (Nemet & Kammen, 2007).

121 **It would increase funding to R&D:** Compared to the total public R&D in renewables and conservation of about $2 billion (IEA, 2007).

121 **This money should be spent:** (Sarewitz & Pielke, 2007:13).

122 **Preliminary studies indicate:** (Kammen & Nemet, 2005; IPCC, 2001a:185) estimate preindustrial CO$_2$ at 280 ppm, thus a CO$_2$ concentration of 560 ppm would probably cause about a 2.38°C rise (A1T from 2000 with 575 ppm [IPCC, 2001a:808, 824]).

122 **The Apollo moon program, with a total:** (Jaffe, Fogarty, & Banks, 1998; Nemet & Kammen, 2007:752; O'Rangers, 2005).

123 **Climate change is most commonly:** (Ereaut & Segnit, 2006:7).

124 **The IPPR points out that "alarmism:** (Ereaut & Segnit, 2006:7).

124 **After the Inquisition's eradication of the actual:** (Behringer, 1999; Oster, 2004).

125 **The pope in 1484 recognized that witches:** (Oster, 2004:217).

125 **As many as half a million individuals:** From 1520 to 1770 (Oster, 2004:220).

125 **Even today, such a climate link is:** (Miguel, 2005).

125 **Less violently, the wet:** (von Storch & Stehr, 2006:108).

125 **The wet summers of the 1910s and:** (von Storch & Stehr, 2006:109).

125 **In 1912, the *Los Angeles Times*:** *Los Angeles Times,* October 7, 1912, "Fifth Ice Age Is on the Way."

125 **In 1923, the *Chicago Tribune* declared on:** (Anderson & Gainor, 2006:9).

126 **However, the world was:** (von Storch & Stehr, 2006:109).

126 **In 1952, *The New York Times*:** The New York Times, August 10, 1952, "Our Changing Climate."

126 **In 1959, it pointed out that glaciers:** *The New York Times,* February 15, 1959, "A Warmer Earth Evident at Poles."

126 **In 1969, it quoted a:** *The New York Times,* February 20, 1969, "Expert Says Arctic Ocean Will Soon Be Open Sea."

126 **Since the 1940s winters have become subtly:** Ponte quoted in (von Storch & Stehr, 2006:109).

126 **Growing glaciers were now seen as a:** (Anderson & Gainor, 2006:13).

126 **It was estimated that the cooling had:** (Anderson & Gainor, 2006:7).

127 *Science Digest* **pointed:** Quoted in (Bray, 1991:82).

127 **The magazine told us that we may be:** *Science News,* March 1, 1975.

127 **Other commentators worried about "worsening:** Nigel Calder quoted in (Bray, 1991:83).

127 *The New York Times* **ran a story headlined:** *The New York Times,* May 21, 1975, "Scientists Ponder Why World's Climate Is Changing."

128 **A collection of green and development groups:** (Simms, Magrath, & Reid, 2004:18).

128 **Leading British commentator George** (Monbiot, 2006:90).

129 **The EU commissioner for the environment:** (EurActiv, 2007).

129 *Time* **tells us, "Be worried:** Cover text of (Kluger, 2006).

129 *New Scientist* **tells us that we are:** (*New Scientist* Anon., 2005).

129 **Perhaps taking the accolades:** Pearson in *Take a Break,* quoted in (Ereaut & Segnit, 2006:30). Perhaps ironically, the magazine's slogan is "Take a Break Magazine—the World Can Wait."

129 **Some leading scientists are beginning:** (Berger, 2007).

129 **One climate scientist has even wondered whether some:** (Berger, 2007).

129 **Crucially, one of the:** (Hulme, 2006).

132 **When the costs of addressing:** (Brahic, 2007).

132 **This would cut U.S. emissions to about:** Using (Nordhaus, 2006d), compared to business as usual.

132 **Yet since the United States will:** From 2.52 to 2.43°C.

132 **Similarly, George Monbiot:** (Monbiot, 2006:3–15).

133 **However, Monbiot says that:** With specific reference to me (Monbiot, 2006:49–53). He also says that he could find lots of other arguments against me, but unfortunately he chose not to produce them.

134 **Should a steward be sacrificed every:** (Monbiot, 2006:175).

134 **Since flying makes up:** (IPCC, 1999a:SPM 4.8), and this includes best estimates for contrails.

134 **As the full Kyoto Protocol would cut:** Kyoto in 2050 would be 5.2 percent below business-as-usual CO_2 emissions (Wigley, 1998:2286).

134 **According to Monbiot, insisting:** (Monbiot, 2006:50).

135 **In the latest review, the previous research:** (Stern, 2006:298). This is similar to the conclusion from a meeting of all economic modelers: "Current assessments determine that the 'optimal' policy calls for a relatively modest level of control of CO_2" (Nordhaus, 1998:18).

135 **This was the state-of-the-art:** For example, (Gibbon, 2006; Stern, 2006; Timmons, 2006). The British UN counselor said worldwide attention has gone "beyond the wildest expectations" of the British government (Hagen, 2007).

135 **predicted apocalyptic:** (Timmons, 2006).

135 **First, Stern finds that the overall:** (Stern, 2006:vi).

135 **The report itself stresses that this is:** (Stern, 2006:vi).

136 **Second, strong action against global warming:** (Stern, 2006:vi).

136 **Virtually everyone has come away:** Even former prime minister Tony Blair: "Stern shows that if we fail to act, the cost of tackling the disruption to people and economies would cost at least five per cent—and possibly as much as 20%—of the world's output. In contrast, the cost of action to halt and reverse climate change would cost just 1%. Or put another way for every £1 we invest now, we can save at least £5 and possibly much more" (Blair, 2006).

136 **The British environment minister pointed out:** (Grice, 2006).

136 **Yet a raft of:** (Byatt et al., 2006; Carter et al., 2006; Dasgupta, 2006; Mendelsohn, 2007; Nordhaus, 2006e; Tol, 2006; Tol & Yohe, 2006; Varian, 2006; Weitzman, 2007; Yohe, 2006).

136 **The review's presentation of the science is:** (Carter et al., 2006:193).

136 **The Review fails to present an:** (Carter et al., 2006:194, 189), compare (Tol & Yohe, 2006:236).

136 **Since the review clearly has its expertise:** In a critical BBC interview, Stern made the following statement, which is hard not to interpret as Stern thinking himself smarter than the IPCC: "We've drawn on the basic science. We have not tried to do new scientific research. We're not scientists."

136 **As several peer-reviewed papers point:** (Byatt et al., 2006:203; Tol & Yohe, 2006:235).

136 **How can it then find conclusions that:** (Byatt et al., 2006:204–5; Tol, 2006:979; Tol & Yohe, 2006:238).

137 **At the same time, the review:** (Tol, 2006:979; Tol & Yohe, 2006:238).

137 **Oddly, it forgets to use this:** They simply stop counting the cost after 2050, while the cost escalates from 2.2 to 6.4 percent of GDP in 2100 (Tol & Yohe, 2006:239).

137 **it would suggest that:** (Dasgupta, 2006).

137 **The costs of action are vastly underestimated:** (Byatt et al., 2006:206).

137 **It implausibly expects costs:** (Mendelsohn, 2007:45).

137 **At the same time, Stern forgets to:** (Tol & Yohe, 2006:239).

138 **Moreover, Stern is not so careful:** (Tol, 2006:979–80).

138 **The most well-known:** (Nordhaus, 2006e:5).

138 *Nature* **tells us that the British government:** (Giles, 2006).

138 **Mike Hulme points out that "this is:** (Giles, 2006).

138 **But the Stern review does not change:** (Stern, 2006:298). This is similar to the conclusion from a meeting of all economic modelers: "Current assessments determine that the 'optimal' policy calls for a relatively modest level of control of CO_2" (Nordhaus, 1998:18).

138 **The IPCC has as its stated goal:** (IPCC, 1999b; IPCC, 2004).

138 **Yet its chairman, R. K. Pachauri:** (Lean, 2005). See also his remarkable introduction to the 2007 IPCC report: "I hope this report will shock people, governments into taking more serious action as you really can't get a more authentic and a more credible piece of scientific work" (Bhalla, 2007). Setting aside the jarring disconnect between serious science and shock value, it is clear that policy neutrality does not entail scaring people senseless.

139 **In the words of several climate scientists:** (von Storch, Stehr, & Ungar, 2004). At least one prominent (unnamed) scientist wanted the Medieval Warm Period to go away, sending an e-mail to what he thought was a fellow believer: "We have to get rid of the Medieval Warm Period" (Deming, 2005).

139 **scientists who dissent:** (Lindzen, 2006).

139 **In April 2000, the:** (Kerr, 2000).

139 **The October 2000 draft:** Preliminary version of (IPCC, 2001a:SPM5).

140 **Yet in the official summary:** (IPCC, 2001a:xi).

140 **When asked about the scientific:** (Pearce, 2001).

140 **Climate policy is here used as a:** "Alternative Development Pathways" is the title in (IPCC, 2001c:95).

140 **Likewise, it is suggested that in:** (IPCC, 2001c:102).

141 **The problem is that "the:** (IPCC, 2001c:101).

141 **Consequently, the IPCC finds that the media:** (IPCC, 2001c:369).

142 **We need to get some broad:** (Schell, 1989).

142 **This is perhaps no more:** See (O'Neill, 2006a) for an extended discussion and following part of his argument here.

142 **This phrase is typically used as a:** Some make a distinction between the science and the policy, but many do not; see (Bailey & English, 2006; Rising Tide, 2007).

142 **David Irving is under:** (Kingston, 2005).

143 **Mark Lynas is an author:** (Lynas, 2004).

143 **He finds that climate denial is:** (Lynas, 2006).

143 **Likewise, David Roberts from** *Grist:* (Roberts, 2006; Roberts, 2006b).

143 **When presented with my economic analyses of:** (Dohm & From, 2004): " 'If you should follow the thinking of Lomborg, then possibly what Hitler did was right' is the harsh commentary from Pachauri."

143 **Fifteen per cent of the:** (Ben-Ami, 2006). This was also Gore's answer on Danish TV when confronted with a critical question about me.

143 **When Gore was challenged on** *The Oprah:* (Winfrey, 2006:slide 13).

144 **Atmospheric physicist John Latham in:** (Blackman, 2006).

144 **Perhaps most important, it could potentially:** Forty billion pounds until CO_2 doubles, or about $2.4 billion annually at 3 percent (Blackman, 2006).

144 **The group denies that it is being dismissive:** (Blackman, 2006).

144 **Although Latham's research has been:** (Bower et al., 2006; Latham, 1990).

145 **At a recent climate demonstration in London:** (O'Neill, 2006b). It is rather hard to imagine this chant feeling particularly rhythmic.

146 **One online editor has compiled:** (Brignell, 2006).

146 **Likewise, the California equivalent of Kyoto:** (California Assembly, 2006). The average Kyoto restriction is 5.2 percent below 1990 levels in 2008–12, whereas California will reduce its to 1990 levels by 2020, starting with some caps in 2012.

146 **Since 1997, British CO_2:** From 1997 to 2004 (EIA, 2006b).

146 **Thus, many in the United Kingdom were:** (Forster, 2006:15).

147 **Perhaps even more clearly, politicians will talk:** (Forster, 2006:12).

147 **Yet at the same time:** (Forster, 2006:13).

148 **Yet when these taxes were actually:** (Forster, 2006:15).

148 **Stopping, or even significantly:** (Tol & Yohe, 2006:245).

CHAPTER 5

150 **Since the Labour government in 1997:** Labour has called for 20 percent CO_2-emissions cuts from 1990 in 2010 in three election manifestos (BBC Anon., 2006a); this translates into a 14.6 percent reduction from 1997 levels. From 1997 to 2004, CO_2 emissions increased 3.4 percent (EIA, 2006b).

150 **Emissions even during the Clinton-Gore:** (EIA, 2006b).

151 **Many believe that dramatic political action will:** Take, for instance, both Gore's "We have to find a way to communicate the direness of the situation" and Hansen's "Scientists have not done a good job communicating with the public"(Fischer, 2006).

151 **Yet the development in EU and:** (EIA, 2006b).

151 **I think if we are going:** (Clinton Global Initiative, 2005:15).

151 **Similarly, one of the:** (Tol, 2007:430).

152 **But before we scoff at 5 percent:** There are many advantages to taxes over emissions caps, mainly that with taxes authorities have an interest in collecting them (because it funds the government), whereas with caps individual countries have much less interest because the benefits are dispersed (global) yet the damages are localized.

153 **When we talk about schools, we:** (Akerhielm, 1995; Angrist & Lavy, 1999; Graddy & Stevens, 2005). Of course, this could be modified in many ways: better-paid teachers, more resources for books, computers, and so on. It is also important to say that more teachers will at least not make schools worse and will likely make them better, as most studies show some or no effect of extra resources, but very few show negative results.

153 **When we talk about hospitals, we:** For example (Fleitas et al., 2006; Gebhardt & Norris, 2006). On the other hand, it is less clear that (after a certain limit) more doctors and bed space is the answer, since they may just make for more visits and more possibilities for infections and harm (Weinberger, Oddone, & Henderson, 1996; Wennberg et al., 2004).

153 **In the United States, 42,600:** (USCB, 2006:672).

153 **Globally, an estimated 1.2:** (Lopez et al., 2006:1751; WHO, 2002:72; WHO, 2004b:3, 172).

154 **Moreover, car traffic also has a:** (AWEA, 2007:2).

154 **About 2 percent of all:** (WHO, 2004b:172).

154 **The total cost is a phenomenal:** (WHO, 2004b:5).

154 **Due to increasing traffic especially in the:** (WHO, 2002:129) puts it second, whereas (WHO, 2004b:5) puts it third.

154 **The answer is simply lowering:** Curiously—and with humor—5 mph is about Monbiot's 96 percent reduction of 55 mph.

155 **Fossil fuels give us low-cost:** This looks only at the marginal benefit of fossil fuels—which is the relevant one for our discussion. On a basic level, though, it is important to remember that they have fundamentally changed our lives. Before fossil fuels, we spent hours gathering wood, contributing to deforestation and soil erosion—as billions in the third world still do (Kammen, 1995). We have electric washing machines that have cut especially women's work dramatically. The historical economist Stanley Lebergott wrote only semi-jokingly: "From 1620 to 1920 the American washing machine was a housewife" (Lebergott, 1993:112). In 1900, a housewife spent seven hours per week laundering, carrying two hundred gallons of water into the house and using a scrub board. Today, she spends eighty-

four minutes with much less strain (Robinson & Godbey, 1997:327). We have a refrigerator that has given us more spare time, allowed us to more easily avoid rotten food, and enabled a healthier diet of fruits and vegetables (Lebergott, 1995:155). By the end of the nineteenth century, human labor made up 94 percent of all industrial work in the United States. Today, it constitutes only 8 percent (Berry, Conkling, & Ray, 1993:131). If we think for a moment of the energy we use in terms of "servants," each with the same work power as a human being, each person in western Europe has access to 150 servants, each in the US about three hundred, and even in India each person has fifteen servants to help (Craig, Vaughan, & Skinner, 1996:103).

155 **Electrical air-conditioning means that people in:** (Davis et al., 2003).

155 **Cheaper fuels would have avoided the deaths of:** Steve Jones of Help the Aged said: "Many pensioners still agonise about whether or not to heat their homes in the cold weather. In the world's fourth richest country, this is simply shameful" (BBC Anon., 2006b).

155 **Fossil fuels allow food to be grown:** The World Cancer Research Fund study estimates that increasing the intake of fruits and vegetables from an average of about 250 grams per day to 400 grams per day would reduce the overall frequency of cancer by around 23 percent (WCRF, 1997:540).

155 **Cars give us an opportunity to get:** (Schäfer, 2006).

156 **About 1.6 billion people don't have access to:** (IEA, 2004:338–40).

156 **Two and a half billion people use:** (IEA, 2006b:419ff).

156 **For many Indian women, searching for:** (IEA, 2006b:428; Kammen, 1995; Kelkar, 2006).

156 **the cost of $1.5 billion annually would:** Mainly in fewer deaths and less time use (IEA, 2006b:440).

156 **Yet studies also show that lowering:** (WHO, 2002:72; WHO, 2004b: 172).

156 **Speed limits across Europe:** See http://www.europe.org/speed limits.html.

158 **Often people will then claim:** (Boykoff & Boykoff, 2006).

158 **Although as a legal principle it really:** (UNCED, 1992:15): "Where there are threats of serious or irreversible damage, lack of full scientific certainty shall not be used as a reason for postponing cost-effective measures to prevent environmental degradation." See also (SEHN, 2007).

158 **We are then back to discussing the:** (Goklany, 2000).

160 **If we focus too much on global:** Going up 67 percent from $227 tril-

lion to $381 trillion in 2100 with the same population (Nakicenovic & IPCC WG III, 2000).

162 **3.5 million lives saved ($7b):** The costs and benefits are from throughout this book; the periods are somewhat different given the diverse nature of the subjects.

163 **When people spend five:** The low-cost estimates at (EcoBusiness-Links, 2007). The average is about seventeen dollars. The benefit estimates are from (Tol, 2005).

Literature

Akbari, H., Pomerantz, M., & Taha, H. (2001). Cool Surfaces and Shade Trees to Reduce Energy Use and Improve Air Quality in Urban Areas. *Solar Energy, 70*(3), 295–310.

Akerhielm, K. (1995). Does Class Size Matter? *Economics of Education Review, 14*(3), 229–241.

Alexander, L. V., Zhang, X., Peterson, T. C., Caesar, J., Gleason, B., Tank, A., et al. (2006). Global Observed Changes in Daily Climate Extremes of Temperature and Precipitation. *Journal of Geophysical Research–Atmospheres, 111*(D5).

Almeida, A. d., Fonseca, P., Schlomann, B., Feilberg, N., & Ferreira, C. (2006). Residential Monitoring to Decrease Energy Use and Carbon Emissions in Europe. Working paper. Retrieved 21-11-06, from http://mail.mtprog.com/CD_Layout/Day_2_22.06.06/1400-1545/ID170_Almeida_final.pdf.

Amstrup, S. C., Durner, G., York, G., Regehr, E., Simac, K., & Douglas, D. (2006, July 14). Polar Bears: Sentinel Species for Climate Change. *Environmental Science Seminar Series*. Retrieved 7-11-06, from http://www.ametsoc.org/atmospolicy/documents/AmstrupFinal.pdf.

Anderson, J. B., Shipp, S. S., Lowe, A. L., Wellner, J. S., & Mosola, A. B. (2002). The Antarctic Ice Sheet during the Last Glacial Maximum and Its Subsequent Retreat History: A Review. *Quaternary Science Reviews, 21*(1–3), 49–70.

Anderson, R. W., & Gainor, D. (2006). Fire and Ice. *Business and Media Institute*. Retrieved 21-1-07, from http://www.businessandmedia.org/specialreports/2006/fireandice/FireandIce.pdf.

Angrist, J. D., & Lavy, V. (1999). Using Maimonides' Rule to Estimate the Effect of Class Size on Scholastic Achievement. *Quarterly Journal of Economics, 114*(2), 533–575.

Annan, K. (2006, November 15). UN Secretary-General Kofi Annan's Address to the Climate Change Conference, as Delivered in Nairobi.

United Nations. Retrieved 2-1-07, from http://www.un.org/News/Press/docs/2006/sgsm10739.doc.htm.

Anon. (2004, November 6). Kyoto Ratification, editorial. *The Washington Post.* Retrieved 18-11-06, from http://www.washingtonpost.com/wp-dyn/articles/A29459-2004Nov5.html.

Anon. (2006, October 28). Urban Heat Island. *New Scientist,* 58.

AP. (2006a, September 26). AP Interview: Schwarzenegger Says Global Warming a Top Priority. Retrieved 6-11-06, from http://www.iht.com/articles/ap/2006/09/26/america/NA_GEN_US_Schwarzenegger_Global_Warming.php.

AP. (2006b, February 23). Japan Tries Some Conservation the Hard Way: Environment Ministry Shuts Off Heating in Race to Meet Kyoto Target. Retrieved 20-11-06, from http://www.msnbc.msn.com/id/11522280/.

Arnell, N. W. (2004). Climate Change and Global Water Resources: SRES Emissions and Socio-economic Scenarios. *Global Environmental Change, 14*(1), 31–52.

Arnell, N. W., Cannell, M. G. R., Hulme, M., Kovats, R. S., Mitchell, J. F. B., Nicholls, R. J., et al. (2002). The Consequences of CO_2 Stabilisation for the Impacts of Climate Change. *Climatic Change, 53*(4), 413–446.

Arnfield, A. J. (2003). Two Decades of Urban Climate Research: A Review of Turbulence, Exchanges of Energy and Water, and the Urban Heat Island. *International Journal of Climatology, 23*(1), 1–26.

Ashton, P. J. (2002). Avoiding Conflicts over Africa's Water Resources. *Ambio, 31*(3), 236–242.

Association of British Insurers. (2005). Financial Risks of Climate Change, Summary Report, by Climate Risk Management. Retrieved 20-12-06, from http://www.abi.org.uk/Display/File/Child/552/Financial_Risks_of_Climate_Change.pdf.

Australian Government Antarctic Division. (2003). Where Do Emperor Penguins Breed? Australian Antarctic Division. Retrieved 15-12-06, from http://www.aad.gov.au/default.asp?casid=2879.

AWEA. (2007). Facts about Wind Energy and Birds. Retrieved 30-1-07, from http://www.awea.org/pubs/factsheets/avianfs.pdf.

BA. (2006). Click for the Climate! *The BA National Science Week 2006.* Retrieved 20-11-06, from http://www.the-ba.net/the-ba/Events/NSEW/AboutNSEW/NSEW_archive/NationalScienceWeek2006/ClimateChange/_ClickfortheClimate.htm.

Bailey, J., & English, O. (2006). Emptying the Sceptic Tank. *Corporate Watch.* Retrieved 26-1-07, from http://www.corporatewatch.org.uk/?lid=2715.

Barber, D. C., Dyke, A., Hillaire-Marcel, C., Jennings, A. E., Andrews, J. T., Kerwin, M. W., et al. (1999). Forcing of the Cold Event of 8,200 Years Ago by Catastrophic Drainage of Laurentide Lakes. *Nature, 400*(6742), 344–348.

Barbraud, C., & Weimerskirch, H. (2001). Emperor Penguins and Climate Change. *Nature, 411*(6834), 183–186.

Barnett, T., Zwiers, F., Hegerl, G., Allen, M., Crowley, T., Gillett, N., et al. (2005). Detecting and Attributing External Influences on the Climate System: A Review of Recent Advances. *Journal of Climate, 18*(9), 1291–1314.

Barnett, T. P., Adam, J. C., & Lettenmaier, D. P. (2005). Potential Impacts of a Warming Climate on Water Availability in Snow-dominated Regions. *Nature, 438*(7066), 303–309.

Basu, R., & Samet, J. M. (2002). Relation between Elevated Ambient Temperature and Mortality: A Review of the Epidemiologic Evidence. *Epidemiologic Reviews, 24*(2), 190–202.

BBC Anon. (2005, January 30). Climate Change "Disaster by 2026." Retrieved 7-11-06, from http://news.bbc.co.uk/1/hi/england/oxfordshire/4218441.stm.

BBC Anon. (2006a, March 28). UK to Miss CO_2 Emissions Target. Retrieved 29-1-07, from http://news.bbc.co.uk/2/hi/science/nature/4849672.stm.

BBC Anon. (2006b, October 27). "Winter Death Toll" Drops by 19%. Retrieved 13-11-06, from http://news.bbc.co.uk/2/hi/uk_news/6090492.stm.

BEA. (2006b). Table 1.1: Current-cost Net Stock of Fixed Assets and Consumer Durable Goods. Bureau of Economic Analysis. Retrieved 23-12-06, from http://www.bea.gov/bea/dn/FA2004/TableView.asp#Mid.

BEA. (2006a). Table 1.1.9: Implicit Price Deflators for Gross Domestic Product. Bureau of Economic Analysis. Retrieved 22-11-06, from http://bea.gov/bea/dn/nipaweb/TableView.asp#Mid.

Beard, J. (2006). DDT and Human Health. *Science of the Total Environment, 355*(1–3), 78–89.

Behringer, W. (1999). Climatic Change and Witch-hunting: The Impact of the Little Ice Age on Mentalities. *Climatic Change, 43*(1), 335–351.

Beinecke, F. (2005). The Natural Resources Defense Council. StopGlobal Warming. Retrieved 29-1-07, from http://www.stopglobalwarming.org/sgw_partner.asp?376.

Ben-Ami, D. (2006, September 18). Global Warming: Time for a Heated Debate. Spiked. Retrieved 26-1-07, from http://www.spiked-online.com/index.php?/site/article/1675/.

Berger, E. (2007, January 22). Climate Scientists Feeling the Heat. *Houston Chronicle.*

Berner, J., Symon, C., Arris, L., Heal, O. W., Arctic Climate Impact Assessment, National Science Foundation (U.S.), et al. (2005). *Arctic Climate Impact Assessment*. New York: Cambridge University Press. Retrieved 7-11-06, from http://www.acia.uaf.edu/pages#scientific.html.

Berry, B. J. L., Conkling, E. C., & Ray, D. M. (1993). *The Global Economy: Resource Use, Locational Choice, and International Trade*. Englewood Cliffs, N.J.: Prentice-Hall.

Bhalla, N. (2007, January 25). U.N. Climate Report Will Shock the World—Chairman. Reuters. Retrieved 26-1-07, from http://www.alertnet.org/thenews/newsdesk/DEL33627.htm.

Bindschadler, R. (2006). The Environment and Evolution of the West Antarctic Ice Sheet: Setting the Stage. *Philosophical Transactions of the Royal Society A: Mathematical, Physical and Engineering Sciences, 364*(1844), 1583–1605. Available at http://dx.doi.org/10.1098/rsta.2006.1790.

BirdLife International. (2004). Aptenodytes forsteri, in: IUCN, *2006 IUCN Red List of Threatened Species*. Retrieved 16-12-06, from http://www.iucnredlist.org/search/details.php/49667/all.

Blackman, S. (2006, November 15). Every Silver Lining Has a Cloud. Spiked. Retrieved 28-1-07, from http://www.spiked-online.com/index.php?/site/article/2097/.

Blair, T. (2004a, September 14). PM Speech on Climate Change. Retrieved 18-11-06, from http://www.pm.gov.uk/output/Page6333.asp.

Blair, T. (2004b, April 27). Speech by the Prime Minister at the Launch of the Climate Group. Retrieved 6-11-06, from http://www.number-10.gov.uk/output/Page5716.asp.

Blair, T. (2006, October 30). PM's Comments at Launch of Stern Review. Retrieved 29-12-06, from http://www.number-10.gov.uk/output/Page10300.asp.

Blakely, S. (1998). Climate Treaty Faces Cold Reception in Congress. *Nation's Business, 86*(2), 8–9.

Bohm, R. (1998). Urban Bias in Temperature Time Series—a Case Study for the City of Vienna, Austria. *Climatic Change, 38*(1), 113–128.

Bohringer, C., & Loschel, A. (2005). Climate Policy beyond Kyoto: Quo Vadis? A Computable General Equilibrium Analysis Based on Expert Judgments. *Kyklos, 58*(4), 467–493.

Bohringer, C., & Vogt, C. (2003). Economic and Environmental Impacts of the Kyoto Protocol. *Canadian Journal of Economics–Revue Canadienne d'Économique, 36*(2), 475–494.

Bosello, F., Roson, R., & Tol, R. S. J. (2006). Economy-wide Estimates of the Implications of Climate Change: Human Health. *Ecological Economics, 58*(3), 579–591.

Bower, K., Choularton, T., Latham, J., Sahraei, J., & Salter, S. (2006). Computational Assessment of a Proposed Technique for Global Warming Mitigation via Albedo-enhancement of Marine Stratocumulus Clouds. *Atmospheric Research, 82*(1–2), 328–336.

Boyd, R. T. (1975). Another Look at the "Fever and Ague" of Western Oregon. *Ethnohistory, 22*(2), 135–154.

Boykoff, J., & Boykoff, M. (2006, July 6). An Inconvenient Principle. CommonDreams.org. Retrieved 30-1-07, from http://www.common dreams.org/views06/0706-26.htm.

Brahic, Catherine. (2007). Costs of Stabilising Global Warming Negligible." *New Scientist.* Retrieved 9-5-07, from http://environment.newscientist .com/article/dn11795-costs-of-stabilising-global-warming-negligible .html.

Bray, A. J. (1991). The Ice-age Cometh—Remembering the Scare of Global Cooling. *Policy Review, 58,* 82–84.

Brazdil, R., Kundzewicz, Z. W., & Benito, G. (2006). Historical Hydrology for Studying Flood Risk in Europe. *Hydrological Sciences Journal– Journal des Sciences Hydrologiques, 51*(5), 739–764.

Brazel, A., Selover, N., Vose, R., & Heisler, G. (2000). The Tale of Two Climates—Baltimore and Phoenix Urban LTER Sites. *Climate Research, 15*(2), 123–135.

Breman, J. G. (2001). The Ears of the Hippopotamus: Manifestations, Determinants, and Estimates of the Malaria Burden. *American Journal of Tropical Medicine and Hygiene, 64*(1–2), 1–11.

Brierly, W. B. (1944). Malaria and Socio-economic Conditions in Mississippi. *Social Forces, 23*(1), 451–459.

Brignell, J. (2006, November 2). Got a Problem? Blame Global Warming! Spiked. Retrieved 27-1-07, from http://www.spiked-online.com/index .php?/site/article/2045/.

Bronstert, A. (2003). Floods and Climate Change: Interactions and Impacts. *Risk Analysis, 23*(3), 545–557. Available at http://www .blackwell-synergy.com/doi/abs/10.1111/1539-6924.00335.

Bryden, H. L., Longworth, H. R., & Cunningham, S. A. (2005). Slowing of the Atlantic Meridional Overturning Circulation at 25 Degrees N. *Nature, 438*(7068), 655–657.

Buncombe, A. (2005, August 19). Climate Change: Will You Listen Now, America? *The Independent.* Retrieved 6-11-06, from http://www .findarticles.com/p/articles/mi_qn4158/is_20050819/ai_n14918176.

Bunting, M. (2006, November 6). It's Hard to Explain, Tom, Why We Did So Little to Stop Global Warming. *The Guardian.* Retrieved 6-11-06, from http://www.guardian.co.uk/commentisfree/story/0,,1940384,00 .html.

Burroughs, W. J. (1997). *Does the Weather Really Matter? The Social Implications of Climate Change.* Cambridge: Cambridge University Press.

Byatt, I., Castles, I., Goklany, I. M., Henderson, D., Lawson, N., McKitrick, R., et al. (2006). The Stern Review: A Dual Critique, Part II: Economic Aspects. *World Economics, 7*(4), 199–232.

Cagle, A., Hubbard, R. (2005). Cold-related Mortality in King County, Washington, U.S.A., 1980–2001. *Annals of Human Biology,* 32(4), 525–537.

California Assembly. (2006, September 27). California Global Warming Solutions Act of 2006, AB 32. Retrieved 27-1-07, from http://www .leginfo.ca.gov/pub/05-06/bill/asm/ab_0001-0050/ab_32_bill_20060927 _chaptered.pdf.

Calvin, W. H. (1998). The Great Climate Flip-flop. *Atlantic Monthly, 281*(1), 47–64.

Campbell-Lendrum, D. H., Corvalán, C. F., & Prüss-Ustün, A. (2003). How Much Disease Could Climate Change Cause? Pp. 133–158 in McMichael et al., 2003.

Carter, R. M., de Freitas, C. R., Goklany, I. M., Holland, D., & Lindzen, R. S. (2006). The Stern Review: A Dual Critique, Part I: The Science. *World Economics, 7*(4), 167–198.

Cazenave, A. (2006). How Fast Are the Ice Sheets Melting? *Science, 314*(5803), 1250–1252.

CDC. (1999). Control of Infectious Diseases, 1900–1999. *JAMA, 282*(11), 1029–1032. Available at http://jama.ama-assn.org.

CDC. (2004). Eradication of Malaria in the United States (1947–1951). Retrieved 30-12-06, from http://0-www.cdc.gov.mill1.sjlibrary.org/ malaria/history/eradication_us.htm.

CDC. (2006). 2006 West Nile Virus Activity in the United States (Reported to CDC as of December 11, 2006). Retrieved 29-12-06, from http://www .cdc.gov/ncidod/dvbid/westnile/surv&controlCaseCount06_detailed .htm.

CEP. (2006). Boosting Innovation and Productivity Growth in Europe: The Hope and the Realities of the EU's "Lisbon Agenda." Centre for Economic Performance. Retrieved 15-3-07, from http://cep.lse.ac.uk/ briefings/pa_lisbon_agenda.pdf.

Chapman, W. L., & Walsh, J. E. (2005). A Synthesis of Antarctic Temperatures. Department of Atmospheric Sciences at the University of Illinois. Retrieved 15-12-06, from http://arctic.atmos.uiuc.edu/ Antarctic.paper.chapwalsh.2005.pdf.

Chase, T. N., Wolter, K., Pielke, R. A., Sr., & Rasool, I. (2007). Was the 2003 European Summer Heat Wave Unusual in a Global Context? *Geophysical Research Letters,* forthcoming. Retrieved 13-11-06, from http://

climatesci.colorado.edu/2006/11/06/was-the-2003-european-summer
-heat-wave-unusual-in-a-global-context/.

Chen, J. L., Wilson, C. R., & Tapley, B. D. (2006). Satellite Gravity Measurements Confirm Accelerated Melting of Greenland Ice Sheet. *Science,*
313(5795), 1958–1960.

Chen, L. X., Zhu, W. Q., Zhou, X. J., & Zhou, Z. J. (2003). Characteristics of
the Heat Island Effect in Shanghai and Its Possible Mechanism.
Advances in Atmospheric Sciences, 20(6), 991–1001.

Chicago Council. (2006a, October 11). Global Views 2006: Comparative
Topline Reports. The Chicago Council on Global Affairs. Retrieved
30-11-06, from http://www.thechicagocouncil.org/dynamic_page
.php?id=56.

Chicago Council. (2006b, October 11). The United States and the Rise of
China and India: Results of a 2006 Multination Survey of Public Opinion. The Chicago Council on Global Affairs. Retrieved 30-11-06, from
http://www.thechicagocouncil.org/dynamic_page.php?id=56.

Chung, U., Choi, J., & Yun, J. I. (2004). Urbanization Effect on the
Observed Change in Mean Monthly Temperatures between 1951–1980
and 1971–2000 in Korea. *Climatic Change, 66*(1–2), 127–136.

Chylek, P., Dubey, M. K., & Lesins, G. (2006). Greenland Warming of
1920–1930 and 1995–2005. *Geophysical Research Letters, 33*(11).

CIA. (2006). CIA World Fact Book. Retrieved 17-12-06, from https://
www.cia.gov/cia/publications/factbook/.

Clinton Global Initiative. (2005, September 15). Special Opening Plenary
Session: Perspectives on the Global Challenges of Our Time. Retrieved
29-1-07, from http://attend.clintonglobalinitiative.org/pdf/transcripts/
plenary/cgi_09_15_05_plenary_1.pdf.

Comrie, A. C. (2000). Mapping a Wind-modified Urban Heat Island in
Tucson, Arizona (with Comments on Integrating Research and Under-
graduate Learning). *Bulletin of the American Meteorological Society,*
81(10), 2417–2431.

Congleton, R. D. (2006). The Story of Katrina: New Orleans and the Political Economy of Catastrophe. *Public Choice, 127*(1–2), 5–30.

Connor, S. (2005, December 1). Fears of Big Freeze as Scientists Detect
Slower Gulf Stream. *The Independent.* Retrieved 27-12-06, from
http://news.independent.co.uk/world/science_technology/article
330454.ece.

Cook, A.J., Fox, A.J., Vaughan, D.G., & Ferrigno, J.G. (2005). Retreating
Glacier Fronts on the Antarctic Peninsula Over the Past Half-century.
Science, 308(5721), 541–544.

Copenhagen Consensus. (2006, October 30). A United Nations Perspective. Retrieved 30-11-06, from http://www.copenhagenconsensus.

com/Admin/Public/DWSDownload.aspx?File=Files%2fFiler%2fCC+ UNP%2fCC06_Outcome.pdf.

Coudrain, A., Francou, B., & Kundzewicz, Z. W. (2005). Glacier Shrinkage in the Andes and Consequences for Water Resources. *Hydrological Sciences Journal–Journal des Sciences Hydrologiques, 50*(6), 925–932.

Cowell, A. (2007, March 14). Britain Drafts Laws to Slash Carbon Emissions. *The New York Times*. Retrieved 15-3-07.

Cox, J. D. (2005). *Climate Crash: Abrupt Climate Change and What It Means for Our Future.* Washington, D.C.: Joseph Henry Press.

Cox, S. (2007, January 25). The Investigation. Radio 4, BBC. Retrieved 28-1-07, from http://www.bbc.co.uk/radio/aod/mainframe.shtml? http://www.bbc.co.uk/radio/aod/radio4_aod.shtml?radio4/ theinvestigation.

Craig, J. R., Vaughan, D. J., & Skinner, B. J. (1996). *Resources of the Earth: Origin, Use and Environmental Impact.* Upper Saddle River, N.J.: Prentice Hall.

CRU. (2006). HadCRUT3 Temperature: Global. Climatic Research Unit, University of East Anglia. Retrieved 1-1-07, from http://www.cru.uea. ac.uk/cru/data/temperature/crutem3gl.txt.

Cullen, N. J., Molg, T., Kaser, G., Hussein, K., Steffen, K., & Hardy, D. R. (2006). Kilimanjaro Glaciers: Recent Areal Extent from Satellite Data and New Interpretation of Observed 20th Century Retreat Rates. *Geophysical Research Letters, 33*(16): Art. no. L16502, 2006, August 31.

Dagoumas, A. S., Papagiannis, G. K., & Dokopoulos, P. S. (2006). An Economic Assessment of the Kyoto Protocol Application. *Energy Policy, 34*(1), 26–39.

Dai, A., Wigley, T. M. L., Boville, B. A., Kiehl, J. T., & Buja, L. E. (2001). Climates of the Twentieth and Twenty-first Centuries Simulated by the NCAR Climate System Model. *Journal of Climate, 14*(4), 485–519.

Dai, A. G., Lamb, P. J., Trenberth, K. E., Hulme, M., Jones, P. D., & Xie, P. P. (2004). The Recent Sahel Drought Is Real. *International Journal of Climatology, 24*(11), 1323–1331.

Dalyell, T. (2004). Westminster Diary. *New Scientist, 181*(2439), 49.

Dana, W. (2006, July 13–27). Al Gore 3.0: The Man Who Won the Presidency in 2000 Is Looser and More Outspoken Than Ever. Is His Global-warming Movie a Warm-up for a Third Run at the White House? *Rolling Stone*. Retrieved 29-11-06, from http://www.rollingstone.com/ news/story/10688399/al_gore_30/print.

Dasgupta, P. (2006, November 11). Comments on the Stern Review's Economics of Climate Change. Retrieved 24-1-07, from http://www .econ.cam.ac.uk/faculty/dasgupta/STERN.pdf.

Davis, R. E., Knappenberger, P. C., Michaels, P. J., & Novicoff, W. M. (2003).

Changing Heat-related Mortality in the United States. *Environmental Health Perspectives, 111*(14), 1712–1718.

Davis, R. E., Knappenberger, P. C., Novicoff, W. M., & Michaels, P. J. (2002). Decadal Changes in Heat-related Human Mortality in the Eastern United States. *Climate Research, 22*(2), 175–184.

Demaree, G. R. (2006). The Catastrophic Floods of February 1784 in and around Belgium—a Little Ice Age Event of Frost, Snow, River Ice . . . and Floods. *Hydrological Sciences Journal–Journal des Sciences Hydrologiques, 51*(5), 878–898.

Deming, D. (2005). Global Warming, the Politicization of Science, and Michael Crichton's "State of Fear." *Journal of Scientific Exploration, 19*(2).

Denton, A. (2006, September 11). Interview with Al Gore. *Enough Rope on Australia ABC.* Retrieved 13-1-07, from http://www.abc.net.au/tv/enoughrope/transcripts/s1734175.htm.

Department of Commerce. (1982). *Survey of Current Business: August 1982.* Retrieved 2-1-07, from http://fraser.stlouisfed.org/publications/SCB/1982/issue/1847.

Department of Commerce. (2006). *Survey of Current Business: December 2006.* Retrieved 2-1-07, from http://www.bea.gov/scb/toc/1206cont.htm.

Dillin, J. (2000). Global Cooling—Mini-ice Age. *Christian Science Monitor, 92*(191), 16.

Dohm, K., & From, L. (2004, April 21). FN-chef: Lomborg Tænker som Hitler [UN executive: Lomborg thinks as Hitler]. *Jyllands-Posten.*

Downton, M., Miller, J. Z. B., & Pielke, R. A. (2005a). Data: Reanalysis of U.S. National Weather Service Flood Loss Database. Retrieved 23-12-06, from http://www.flooddamagedata.org/national.html.

Downton, M., Miller, J. Z. B., & Pielke, R. A. (2005b). Reanalysis of U.S. National Weather Service Flood Loss Database. *Natural Hazards Review, 2*(4), 157–166. Retrieved 19-12-06, from http://sciencepolicy.colorado.edu/admin/publication_files/resource-478-2005.16.pdf.

DW staff. (2006, September 28). Merkel to Target Climate Change as G8, EU Leader. *Deutsche Welle.* Retrieved 6-11-06, from http://www.dw-world.de/dw/article/0,2144,2188336,00.html.

Easterling, D. R., Evans, J. L., Groisman, P. Y., Karl, T. R., Kunkel, K. E., & Ambenje, P. (2000). Observed Variability and Trends in Extreme Climate Events: A Brief Review. *Bulletin of the American Meteorological Society, 81*(3), 417–425.

EB. (2006a). Flood Legend. *Encyclopedia Britannica.* Retrieved 9-12-06 from http://www.britannica.com/ebi/article-9274347.

EB. (2006b). Gulf Stream. *Encyclopedia Britannica.* Retrieved 25-12-06 from http://search.eb.com.esc-proxy.cbs.dk/eb/article-9038484.

EB. (2006c). Holcene Epoch. *Encyclopedia Britannica*. Retrieved 7-12-06 from http://search.eb.com.esc-proxy.lib.cbs.dk/eb/article-9117456.

EB. (2006d). Ice Age. *Encyclopedia Britannica*. Retrieved 7-12-06 from http://search.eb.com.esc-proxy.lib.cbs.dk/eb/article-9041958.

Ebi, K. L., Mills, D. M., Smith, J. B., & Grambsch, A. (2006). Climate Change and Human Health Impacts in the United States: An Update on the Results of the US National Assessment. *Environmental Health Perspectives, 114*(9), 1318–1324.

EcoBusinessLinks. (2007). How Much Does Carbon Offsetting Cost? Price Survey! EcoBusinessLinks.com. Retrieved 1-2-07, from http://www.ecobusinesslinks.com/carbon_offset_wind_credits_carbon_reduction.htm.

EDD. (2006a). Miami Beach Statistical Abstract 2000–2006. Economic Development Department. Retrieved 10-12-06, from http://www.miamibeachfl.gov/newcity/depts/econdev/Statistical%20Abstract%20(Long).pdf.

EDD. (2006b). Tourism Overview. Economic Development Department. Retrieved 10-12-06, from http://www.miamibeachfl.gov/newcity/depts/econdev/visitors%20Profile.asp.

Edwards, G. (2006, April 5). Hot in Here. *Rolling Stone,* 27.

EIA. (1999). *Carbon Dioxide Emissions from the Generation of Electric Power in the United States.* U.S. Energy Information Administration. Retrieved 21-11-06, from http://www.eia.doe.gov/cneaf/electricity/page/co2_report/co2emiss99.pdf.

EIA. (2002). *Updated State-level Greenhouse Gas Emission Coefficients for Electricity Generation 1998–2000.* U.S. Energy Information Administration. Retrieved 21-11-06, from http://tonto.eia.doe.gov/FTP-ROOT/environment/e-supdoc-u.pdf.

EIA. (2006a). Emissions of Greenhouse Gases in the United States 2005. U.S. Energy Information Administration. Retrieved 30-11-06, from http://www.eia.doe.gov/oiaf/1605/ggrpt/pdf/057305.pdf.

EIA. (2006b). International Energy Annual 2004. U.S. Energy Information Administration. Retrieved 30-11-06, from http://www.eia.doe.gov/iea/.

EIA. (2006c). *International Energy Outlook 2006.* U.S. Energy Information Administration. Retrieved 20-11-06, from http://www.eia.doe.gov/oiaf/ieo/index.html.

EIA. (2006d). *Monthly Energy Review: October 2006.* U.S. Energy Information Administration.

EIA. (2006e). US Historical CO_2 Emissions. U.S. Energy Information Administration.

EIA. (2006f). Voluntary Reporting of Greenhouse Gases Program: Fuel and Energy Source Codes and Emission Coefficients. U.S. Energy Information Administration. Retrieved 22-11-06, from http://www.eia.doe.gov/oiaf/1605/coefficients.html.

Eilperin, J. (2004). Study Says Polar Bears Could Face Extinction. *The Washington Post.* Retrieved 7-11-06, from http://www.washingtonpost.com/wp-dyn/articles/A35233-2004Nov8.html.

EPA. (2000, April). Average Annual Emissions and Fuel Consumption for Passenger Cars and Light Trucks. U.S. Environmental Protection Agency. Retrieved 21-11-06, from http://www.epa.gov/otaq/consumer/f00013.pdf.

Epstein, P. R. (2000). Is Global Warming Harmful to Health? *Scientific American, 283*(2), 50–57.

Epstein, P. R., Diaz, H. F., Elias, S., Grabherr, G., Graham, N. E., Martens, W. J. M., et al. (1998). Biological and Physical Signs of Climate Change: Focus on Mosquito-borne Diseases. *Bulletin of the American Meteorological Society, 79*(3), 409–417.

Ereaut, G., & Segnit, N. (2006). Warm Words: How Are We Telling the Climate Story and Can We Tell It Better? Institute for Public Policy Research. Retrieved 20-1-07, from http://www.ippr.org.uk/members/download.asp?f=/ecomm/files/warm_words.pdf&a=skip.

EU. (1996a, June 11). 1st UNFCC Communication EN. Commission of the European Communities. Retrieved 27-11-06, from http://ec.europa.eu/environment/climat/pdf/1st_unfcc_communication_en.pdf.

EU. (1996b, June 25). Communication on Community Strategy on Climate Change. 1939th Council meeting, Luxembourg.

EU. (2001, June TK). ECCP Report. Commission of the European Communities. Retrieved 27-11-06, from http://ec.europa.eu/environment/climat/pdf/eccp_report_0106.pdf.

EU. (2005, February 9). Winning the Battle against Global Climate Change. Commission of the European Communities. Retrieved 28-11-06, from http://eur-lex.europa.eu/LexUriServ/site/en/com/2005/com2005_0035en01.pdf.

EU. (2007a). Gross Domestic Expenditure on R&D. *EUROSTAT.* Retrieved 15-3-07, from http://epp.eurostat.ec.europa.eu/portal/page?_pageid=1996,39140985&_dad=portal&_schema=PORTAL&screen=detailref&language=en&product=Yearlies_new_science_technology&root=Yearlies_new_science_technology/I/I1/ir021.

EU. (2007b). Lisbon Strategy. *Europa Glossary.* Retrieved 15-3-07, from http://europa.eu/scadplus/glossary/lisbon_strategy_en.htm.

EU. (2007c, March 9). Presidency Conclusions of the Brussels European Council, 8/9 March 2007. Retrieved 15-3-07, from http://www.consilium.europa.eu/ueDocs/cms_Data/docs/pressData/en/ec/93135.pdf.

EurActiv. (2007, January 12). EU Defends Leadership in "World War" on Climate Change. EurActive.com. Retrieved 22-1-07, from http://www.euractiv.com/en/energy//article-160848.

Evans, E., Ashley, R., Hall, J., Penning-Rowsell, E., Saul, A., Sayers, P., et al. (2004). *Foresight, Future Flooding. Scientific Summary: Volume 1, Future Risks and Their Drivers*. London: Office of Science and Technology. Retrieved 14-1-07, from http://www.foresight.gov.uk/Previous_Projects/Flood_and_Coastal_Defence/Reports_and_Publications/Volume1/Thanks.htm.

Evans, E., Ashley, R., Hall, J., Penning-Rowsell, E., Sayers, P., Thorne, C., et al. (2004). *Foresight, Future Flooding. Scientific Summary: Volume 2, Managing Future Risks*. London: Office of Science and Technology. Retrieved 14-1-07, from http://www.foresight.gov.uk/Previous_Projects/Flood_and_Coastal_Defence/Reports_and_Publications/Volume2/Thanks2.htm.

FAO. (2001). *Global Forest Resources Assessment 2000*. UN Food and Agriculture Organization. Retrieved 25-12-06, from http://www.fao.org/docrep/004/Y1997E/y1997e00.htm#Contents.

FAO. (2006). *World Agriculture: Towards 2030/2050—Interim Report*. UN Food and Agriculture Orgaization. Retrieved 2-1-07, from http://www.fao.org/es/ESD/AT2050web.pdf.

FAO. (2007). FAOSTAT database. Available at http://faostat.fao.org/default.aspx.

Fischer, D. (2006, December 15). Gore Urges Scientists to Speak Up. *Contra Costa Times*. Retrieved 29-1-07, from http://www.truthout.org/cgi-bin/artman/exec/view.cgi/67/24524.

Fischer, G., Shah, M., Tubiello, F. N., & van Velhuizen, H. (2005). Socioeconomic and Climate Change Impacts on Agriculture: An Integrated Assessment, 1990–2080. *Philosophical Transactions of the Royal Society B: Biological Sciences, 360*(1463), 2067–2083.

Fischer, G., Shah, M., & Velthuizen, H. v. (2002). *Climate Change and Agricultural Vulnerability*. International Institute for Applied Systems Analysis for World Summit on Sustainable Development, Johannesburg 2002. Retrieved 3-1-07, from http://www.iiasa.ac.at/Research/LUC/JB-Report.pdf.

Fischer, G., Velthuizen, H. v., Shah, M., & Nachtergaele, F. (2002). *Global Agro-ecological Assessment for Agriculture in the 21st Century: Methodology and Results*. International Institute for Applied Systems Analysis and UN Food and Agriculture Organization. Retrieved 3-1-07, from http://www.iiasa.ac.at/Admin/PUB/Documents/RR-02-002.pdf.

Fleitas, I., Caspani, C. C., Borras, C., Plazas, M. C., Miranda, A. A., Brandar,

M. E., et al. (2006). The Quality of Radiology Services in Five Latin American Countries. *Pan American Journal of Public Health–Revista Panamericana de Salud Pública, 20*(2–3), 113–124.

FOE. (2006). Climate: Climate Change. Friends of the Earth. Retrieved 17-12-06, from http://www.foe.co.uk/campaigns/climate/issues/climate_change/.

Forster, A. (2006, November 30). Can We Go on Building Roads and Runways *and* Save the Planet? LTT Online. Retrieved 28-1-07, from http://www.staff.livjm.ac.uk/spsbpeis/LTT-interviewNo06.pdf.

Fowler, H. J., & Archer, D. R. (2006). Conflicting Signals of Climatic Change in the Upper Indus Basin. *Journal of Climate, 19*(17), 4276–4293.

GAO. (1995). *Midwest Flood: Information on the Performance, Effects, and Control of Levees* (vol. GAO/RCED-95-125). U.S. General Accounting Office. Retrieved 22-12-06, from http://www.gao.gov/cgi-bin/getrpt?RCED-95-125.

Gebhardt, J. G., & Norris, T. E. (2006). Acute Stroke Care at Rural Hospitals in Idaho: Challenges in Expediting Stroke Care. *Journal of Rural Health, 22*(1), 88–91.

Gelbspan, R. (2004). *Boiling Point: How Politicians, Big Oil and Coal, Journalists, and Activists Are Fueling the Climate Crisis—and What We Can Do to Avert Disaster.* New York: Basic Books.

Gelbspan, R. (2005, August 30). Katrina's Real Name. *The Boston Globe.* Retrieved 17-12-06, from http://www.boston.com/news/weather/articles/2005/08/30/katrinas_real_name/.

Giannini, A., Saravanan, R., & Chang, P. (2003). Oceanic Forcing of Sahel Rainfall on Interannual to Interdecadal Time Scales. *Science, 302*(5647), 1027–1030.

Gibbon, G. (2006, October 30). Government Pledges Action. *Channel4 News.* Retrieved 24-1-07, from http://www.channel4.com/news/special-reports/special-reports-storypage.jsp?id=3757.

Giles, J. (2006, November 2). How Much Will It Cost to Save the World? *Nature,* 6–7.

GISS. (2006). Amundsen-Scott Temperature Data, 1957–2006. Goddard Institute for Space Studies.

Goklany, I. M. (2000). Applying the Precautionary Principle to Global Warming. Weidenbaum Center Working Paper no. PS 158. Retrieved 30-1-07, from http://ssrn.com/abstract=250380.

Goklany, I. M. (2006). *The Improving State of the World: Why We're Living Longer, Healthier, More Comfortable Lives on a Cleaner Planet.* Washington, D.C.: Cato Institute.

Golub, A., Markandya, A., & Marcellino, D. (2006). Does the Kyoto Proto-col Cost Too Much and Create Unbreakable Barriers for Economic Growth? *Contemporary Economic Policy, 24*(4), 520–535.

Gore, A. (2006a, November 19). At Stake Is Nothing Less Than the Survival of Human Civilisation. *Sunday Telegraph*. Retrieved 19-12-06, from http://www.telegraph.co.uk/news/main.jhtml;jsessionid=FAFKOMP HJAYNJQFIQMFSFFOAVCBQ0IV0?xml=/news/2006/11/19/nclim19.xml&page=1.

Gore, A. (2006b). *An Inconvenient Truth: The Movie*. Paramount DVD.

Gore, A., & Melcher Media. (2006). *An Inconvenient Truth: The Planetary Emergency of Global Warming and What We Can Do about It*. Emmaus, Pa.: Rodale Press.

Graddy, K., & Stevens, M. (2005). The Impact of School Resources on Stu-dent Performance: A Study of Private Schools in the United Kingdom. *Industrial and Labor Relations Review, 58*(3), 435–451.

Graham-Harrison, E. (2006, October 30). China Hopes for Post-2012 Kyoto Deal within 2 Years. Reuters. Retrieved 19-11-06, from http://www.planetark.com/dailynewsstory.cfm/newsid/38721/story.htm.

Greater London Authority. (2006). *London's Urban Heat Island: A Sum-mary for Decision Makers*. Retrieved 15-11-06, from http://www.london.gov.uk/mayor/environment/climate-change/uhi.jsp.

Greenpeace. (2001). Kilimanjaro Set to Lose Its Ice Field by 2015 Due to Climate Change. News release. Retrieved 7-12-06, from http://www.commondreams.org/news2001/1106-02.htm.

Greenpeace. (2004). Global Warnings. Retrieved 17-12-06, from http://www.greenpeace.org/international/news/extreme-weather-warnings.

Greenpeace. (2006a). Climate Change. Retrieved 19-12-06, from http://www.greenpeace.org/seasia/en/asia-energy-revolution/climate-change.

Greenpeace. (2006b). Sea Level Rise. Retrieved 15-12-06, from http://www.greenpeace.org/international/campaigns/climate-change/impacts/sea_level_rise.

Gregory, J., & Huybrechts, P. (2006). Ice-sheet Contributions to Future Sea-level Change. *Philosophical Transactions of the Royal Society A: Mathematical, Physical and Engineering Sciences, 364*(1844), 1709–1731.

Grice, A. (2006, November 20). Slow Talks Could Leave Climate Deal in "Tatters." *The Independent*. Retrieved 20-11-06, from http://news.independent.co.uk/environment/article1998840.ece.

Grigg, D. B. (1993). *The World Food Problem* (2d ed.). Cambridge: Black-well.

216   *Literature*

Groisman, P. Y., Knight, R. W., Easterling, D. R., Karl, T. R., Hegerl, G. C., & Razuvaev, V. A. N. (2005). Trends in Intense Precipitation in the Climate Record. *Journal of Climate, 18*(9), 1326–1350.

Groisman, P. Y., Knight, R. W., & Karl, T. R. (2001). Heavy Precipitation and High Streamflow in the Contiguous United States: Trends in the Twentieth Century. *Bulletin of the American Meteorological Society, 82*(2), 219–246.

Grubb, M. (2004). Kyoto and the Future of International Climate Change Responses: From Here to Where? *International Review for Environmental Strategies, 5*(1), 15–38.

Grubb, M., Kohler, J., & Anderson, D. (2002). Induced Technical Change in Energy and Environmental Modeling: Analytic Approaches and Policy Implications. *Annual Review of Energy and the Environment, 27*, 271–308.

Grzimek. (n.d.). Emperor Penguin. *Grzimek's Animal Life Encyclopedia.* Retrieved 16-12-06, from http://www.answers.com/topic/emperor-penguin.

Guerra, C. A., Snow, R. W., & Hay, S. I. (2006). Mapping the Global Extent of Malaria in 2005. *Trends in Parasitology, 22*(8), 353–358.

Hagen, J. (2007, January 19). Act on Global Warming Now or Pay Later: The Stern Review. *UN Chronicle Online Edition.* Retrieved 24-1-07, from http://www.un.org/Pubs/chronicle/2007/webArticles/011907_stern.htm.

Hahn, R. W. (1996). *Risks, Costs, and Lives Saved: Getting Better Results from Regulation.* New York: Oxford University Press.

Hanley, C. J. (2006, December 18). Malaria Cases Climb in African Highlands. Associated Press.

Hansen, B., Osterhus, S., Quadfasel, D., & Turrell, W. (2004). Already the Day after Tomorrow? *Science, 305*(5686), 953–954.

Harden, B. (2005, July 7). Experts Predict Polar Bear Decline: Global Warming Is Melting Their Ice Pack Habitat. *The Washington Post.* Retrieved 7-11-06, from http://www.washingtonpost.com/wp-dyn/content/article/2005/07/06/AR2005070601899.html.

Hassol, S. J. (2004). *Impacts of a Warming Arctic : Arctic Climate Impact Assessment.* New York: Cambridge University Press. Retrieved 7-11-06, from http://www.acia.uaf.edu/pages/overview.html.

Hawkins, D. (2001, July 10). Climate Change Technology and Policy Options. Director of NRDC's climate center, comments for U.S. Senate Committee on Commerce, Science, and Transportation. Retrieved 29-1-07, from http://www.nrdc.org/globalWarming/tdh0701.asp.

Hay, S. I., Cox, J., Rogers, D. J., Randolph, S. E., Stern, D. I., Shanks, G. D., et al. (2002a). Climate Change and the Resurgence of Malaria in the East African Highlands. *Nature, 415*(6874), 905–909.

Hay, S. I., Cox, J., Rogers, D. J., Randolph, S. E., Stern, D. I., Shanks, G. D., et al. (2002b). Climate Change—Regional Warming and Malaria Resurgence—Reply. *Nature, 420*(6916), 628.

Hay, S. I., Guerra, C. A., Tatem, A. J., Atkinson, P. M., & Snow, R. W. (2005). Urbanization, Malaria Transmission and Disease Burden in Africa. *Nature Reviews–Microbiology, 3*(1), 81–90.

Hay, S. I., Guerra, C. A., Tatem, A. J., Noor, A. M., & Snow, R. W. (2004). The Global Distribution and Population at Risk of Malaria: Past, Present, and Future. *Lancet Infectious Diseases, 4*(6), 327–336.

Hay, S. I., Rogers, D. J., Randolph, S. E., Stern, D. I., Cox, J., Shanks, G. D., et al. (2002). Hot Topic or Hot Air? Climate Change and Malaria Resurgence in East African Highlands. *Trends in Parasitology, 18*(12), 530–534.

Helm, D. (2003). The Assessment: Climate-change Policy. *Oxford Review of Economic Policy, 19*(3), 349–361.

Henderson, M. (2005, December 1). Britain Faces Big Freeze as Gulf Stream Loses Strength. *The Times.* Retrieved 27-12-06, from http://www.timesonline.co.uk/article/0,,2-1898493,00.html.

Hinkel, K. M., Nelson, F. E., Klene, A. F., & Bell, J. H. (2003). The Urban Heat Island in Winter at Barrow, Alaska. *International Journal of Climatology, 23*(15), 1889–1905.

Hoerling, M., Hurrell, J., Eischeid, J., & Phillips, A. (2006). Detection and Attribution of Twentieth-century Northern and Southern African Rainfall Change. *Journal of Climate, 19*(16), 3989–4008.

Horton, E. B., Folland, C. K., & Parker, D. E. (2001). The Changing Incidence of Extremes in Worldwide and Central England Temperatures to the End of the Twentieth Century. *Climatic Change, 50*(3), 267–295.

Howat, I. M., Joughin, I., & Scambos, T. A. (2007). Rapid Changes in Ice Discharge from Greenland Outlet Glaciers. *Science, 315*(5818), 1559–1561. Available at http://www.sciencemag.org/cgi/content/abstract/315/5818/1559.

Hughes, M. K., & Diaz, H. F. (1994). Was There a Medieval Warm Period, and If So, Where and When? *Climatic Change, 26*(2–3), 109–142.

Hulme, M. (2006, November 4). Chaotic World of Climate Truth. BBC. Retrieved 22-1-07, from http://news.bbc.co.uk/2/hi/science/nature/6115644.stm.

Humlum, O. (n.d.). Antarctic Temperature Changes during the Observational Period. UNIS, Department of Geology, Svalbard, Norway. Retrieved 15-12-06, from http://www.unis.no/research/geology/Geo_research/Ole/AntarcticTemperatureChanges.htm.

Hung, T., Uchihama, D., Ochi, S., & Yasuoka, Y. (2006). Assessment with Satellite Data of the Urban Heat Island Effects in Asian Mega Cities. *International Journal of Applied Earth Observation and Geoinformation, 8*(1), 34–48.

Huntington, T. G. (2006). Evidence for Intensification of the Global Water Cycle: Review and Synthesis. *Journal of Hydrology, 319*(1–4), 83–95.

Hutton, G., & Haller, L. (2004). *Evaluation of the Costs and Benefits of Water and Sanitation Improvements at the Global Level.* WHO/SDE/WSH/04.04: World Health Organization. Retrieved 8-1-07, from http://www.who.int/water_sanitation_health/wsh0404.pdf.

Huybrechts, P., & de Wolde, J. (1999). The Dynamic Response of the Greenland and Antarctic Ice Sheets to Multiple-century Climatic Warming. *Journal of Climate, 12*(8), 2169–2188.

IEA. (2004). *World Energy Outlook 2004.* Paris: International Energy Agency Publications.

IEA. (2006a). *CO_2 Emissions from Fuel Combustion 1971–2004.* Paris: International Energy Agency Publications.

IEA. (2006b). *World Energy Outlook 2006.* Paris: International Energy Agency Publications.

IEA. (2007). IEA Energy Technology R&D Statistics Service. International Energy Agency. Available at http://www.iea.org/rdd/ReportFolders/ReportFolders.aspx?CS_referer=&CS_ChosenLang=en.

Ijumba, J. N., Mosha, F. W., & Lindsay, S. W. (2002). Malaria Transmission Risk Variations Derived from Different Agricultural Practices in an Irrigated Area of Northern Tanzania. *Medical and Veterinary Entomology, 16*(1), 28–38.

IMF. (2006, September). *World Economic Outlook: Financial Systems and Economic Cycles.* International Monetary Fund. Retrieved 20-11-06, from http://www.imf.org/external/pubs/ft/weo/2006/02/index.htm.

Insurance Journal. (2006, April 18). Sound Risk Management, Strong Investment Results Prove Positive for P/C Industry. Retrieved 20-12-06, from http://www.insurancejournal.com/news/national/2006/04/18/67389.htm.

IPCC. (1999a). *Aviation and the Global Atmosphere.* Cambridge: Cambridge University Press. Retrieved 24-1-07, from http://www.grida.no/climate/ipcc/aviation/index.htm.

IPCC. (1999b). Procedures for the Preparation, Review, Acceptance, Adoption, Approval and Publication of IPCC Reports. United Nations. Retrieved 26-1-07, from http://www.climatescience.gov/Library/ipcc/app-a.pdf.

IPCC. (2001a). *Climate Change 2001: WGI: The Scientific Basis. Contribution of Working Group I to the Third Assessment Report of the Intergov-*

ernmental Panel on Climate Change [Houghton, J. Ding, T.,Y., Griggs, D. J., Noguer, M., van der Linden, P .J., Dai, X., Maskell, K., and Johnson, C. A. (eds.)]. Cambridge: Cambridge University Press. Available at http://www.grida.no/climate/ipcc_tar/wg1/index.htm.

IPCC. (2001b). *Climate Change 2001: WGII: Impacts, Adaptation and Vulnerability.* Cambridge: Cambridge University Press. Available at http://www.grida.no/climate/ipcc_tar/wg2/index.htm.

IPCC. (2001c). *Climate Change 2001: WGIII: Mitigation.* Cambridge: Cambridge University Press. Available at http://www.grida.no/climate/ipcc_tar/wg3/index.htm.

IPCC. (2004, June 22). Deputy Secretary. United Nations. Retrieved 26-1-07, from http://notesapps.unon.org/notesapps/vacs.nsf/4c8db34864 91c2f543256c3f004b0401/6e11b57c7b77623243256fe300289728?Open Document.

IPCC. (2007b). *Climate Change 2007: WGI: The Physical Science Basis.* Cambridge: Cambridge University Press.

IPCC. (2007a). *Climate Change 2007: WGI: Summary for Policymakers.* Retrieved 13-2-07, from http://www.ipcc.ch/SPM2feb07.pdf.

IPCC. (2007c). *Climate Change 2007: WGII: Impacts, Adaptation and Vulnerability.* Cambridge: Cambridge University Press.

IPCC. (2007d). *Climate Change 2007: WGII: Summary for Policymakers.* Retrieved 6-4-07, from http://www.ipcc.ch/SPM6avr07.pdf.

IPCC, Bruce, J. P., Yi, H.-s. o., Haites, E. F., & Working Group III. (1996). *Climate Change 1995: Economic and Social Dimensions of Climate Change.* New York: Intergovernmental Panel on Climate Change; Cambridge University Press.

IPCC & Houghton, J. T. (1996). *Climate Change 1995: The Science of Climate Change.* New York: Cambridge University Press.

IPCC, Houghton, J. T., Jenkins, G. J., Ephraums, J. J., & Working Group I. (1990). *Climate Change: The IPCC Scientific Assessment.* New York: Cambridge University Press.

Iredale, W. (2005, December 18). Polar Bears Drown as Ice Shelf Melts. *Sunday Times.* Retrieved 7-11-06, from http://www.timesonline.co.uk/article/0,,2087-1938132,00.html.

IUCN Species Survival Commission. (2001). *Polar Bears: Proceedings of the 13th Working Meeting of the IUCN/SSC Polar Bear Specialist Group, 23–28 June 2001, Nuuk, Greenland.* Retrieved 6-11-2006, from http://pbsg.npolar.no/docs/PBSG13proc.pdf.

Jaffe, A. B., Fogarty, M. S., & Banks, B. A. (1998). Evidence from Patents and Patent Citations on the Impact of NASA and Other Federal Labs on Commercial Innovation. *Journal of Industrial Economics, 46*(2), 183–205.

Jaffe, A. B., Newell, R. G., & Stavins, R. N. (1999). Energy-efficient Technologies and Climate Change Policies: Issues and Evidence. *Resources for the Future, Climate Issue Brief no. 19.* Retrieved 21-11-06, from http://ksghome.harvard.edu/~rstavins/Selected_Articles/RFF_Energy_Effiient_Tech_and_Climate_Change_Policies.pdf.

Jamison, D. T., Feachem, R. G., Makgoba, M. W., Bos, E. R., Baingana, F. K., Hofman, K. J., et al. (2006). *Disease and Mortality in Sub-Saharan Africa.* Washington, D.C.: World Bank.

Jenouvrier, S., Barbraud, C., & Weimerskirch, H. (2006). Sea Ice Affects the Population Dynamics of Adelie Penguins in Terre Adelie. *Polar Biology, 29*(5), 413–423.

Jevrejeva, S., Grinsted, A., Moore, J. C., & Holgate, S. (2006). Nonlinear Trends and Multiyear Cycles in Sea Level Records. *Journal of Geophysical Research–Oceans, 111*(C9).

Joerin, U. E., Stocker, T. F., & Schluchter, C. (2006). Multicentury Glacier Fluctuations in the Swiss Alps during the Holocene. *Holocene, 16*(5), 697–704.

Johannessen, O. M., Khvorostovsky, K., Miles, M. W., & Bobylev, L. P. (2005). Recent Ice-sheet Growth in the Interior of Greenland. *Science, 310*(5750), 1013–1016.

Jones, P. D., Horton, E. B., Folland, C. K., Hulme, M., Parker, D. E., & Basnett, T. A. (1999). The Use of Indices to Identify Changes in Climatic Extremes. *Climatic Change, 42*(1), 131–149.

Jungclaus, J. H., Haak, H., Esch, M., Roeckner, E., & Marotzke, J. (2006). Will Greenland Melting Halt the Thermohaline Circulation? *Geophysical Research Letters, 33*(17).

Kammen, D. M. (1995). Cookstoves for the Developing-world. *Scientific American, 273*(1), 72–75.

Kammen, D. M., & Nemet, G. F. (2005). Reversing the Incredible Shrinking Energy R&D Budget. *Issues in Science and Technology,* Fall, 84–88. Retrieved 20-1-07, from http://rael.berkeley.edu/files/2005/Kammen-Nemet-ShrinkingRD-2005.pdf.

Karl, T. R., & Trenberth, K. E. (1999). The Human Impact on Climate. *Scientific American, 281*(6), 100–105.

Kaser, G., Hardy, D. R., Molg, T., Bradley, R. S., & Hyera, T. M. (2004). Modern Glacier Retreat on Kilimanjaro as Evidence of Climate Change: Observations and Facts. *International Journal of Climatology, 24*(3), 329–339.

Kavuncu, Y. O., & Knabb, S. D. (2005). Stabilizing Greenhouse Gas Emissions: Assessing the Intergenerational Costs and Benefits of the Kyoto Protocol. *Energy Economics, 27*(3), 369–386.

Keatinge, W. R., & Donaldson, G. C. (2004). The Impact of Global Warming on Health and Mortality. *Southern Medical Journal, 97*(11), 1093–1099.

Keatinge, W. R., Donaldson, G. C., Cordioli, E. A., Martinelli, M., Kunst, A. E., Mackenbach, J. P., et al. (2000). Heat Related Mortality in Warm and Cold Regions of Europe: Observational Study. *British Medical Journal, 321*(7262), 670–673.

Kelkar, G. (2006, May 8). The Gender Face of Energy. Presentation at CSD 14 Learning Centre, United Nations. Retrieved 30-1-07, from http://www.un.org/esa/sustdev/csd/csd14/lc/presentation/gender2.pdf.

Kennedy, R. F. (2005, August 29). "For They That Sow the Wind Shall Reap the Whirlwind." The Huffington Post. Retrieved 17-12-06, from http://www.huffingtonpost.com/robert-f-kennedy-jr/for-they-that-sow-the-_b_6396.html.

Kerr, R. A. (2000, April 25). U.N. to Blame Global Warming on Humans. *ScienceNow Daily News*, 1.

Kerr, R. A. (2005). Global Climate Change—the Atlantic Conveyor May Have Slowed, but Don't Panic Yet. *Science, 310*(5753), 1403.

Kerr, R. A. (2006). Global Climate Change—False Alarm: Atlantic Conveyor Belt Hasn't Slowed Down After All. *Science, 314*(5802), 1064.

Kerr, R. A. (2007, February 15). Predicting Fate of Glaciers Proves Slippery Task. *ScienceNow Daily News*, 2. Available at http://sciencenow.sciencemag.org/cgi/content/full/2007/215/2.

Khaleque, V, (2006, December 7). Bangladesh Is Paying a Cruel Price for the West's Excesses. *The Guardian*. Retrieved 1-1-07, from http://environment.guardian.co.uk/climatechange/story/0,,1966012,00.html.

Khandekar, M. L., Murty, T. S., & Chittibabu, P. (2005). The Global Warming Debate: A Review of the State of Science. *Pure and Applied Geophysics, 162*(8–9), 1557–1586.

King, D. A. (2004). Environment—Climate Change Science: Adapt, Mitigate, or Ignore? *Science, 303*(5655), 176–177.

Kingston, M. (2005, November 21). Himalayan Lakes Disaster. The Daily Briefing. Retrieved 26-1-07, from http://webdiary.com.au/cms/?q=node/986.

Kluger, J. (2006, April 3). Polar Ice Caps Are Melting Faster Than Ever . . . More and More Land Is Being Devastated by Drought . . . Rising Waters Are Drowning Low-lying Communities . . . by Any Measure, Earth Is at . . . the Tipping Point. *Time*. Retrieved 6-11-06, from http://www.time.com/time/magazine/article/0,9171,1176980,00.html.

Konradsen, F., van der Hoek, W., Amerasinghe, F. P., Mutero, C., & Boelee, E. (2004). Engineering and Malaria Control: Learning from the Past 100 Years. *Acta Tropica, 89*(2), 99–108.

Kooyman, G. L. (1993). Breeding Habitats of Emperor Penguins in the Western Ross Sea. *Antarctic Science, 5*(2), 143–148.

Krauss, C. (2006, May 27). Bear Hunting Caught in Global Warming Debate. *The New York Times.* Retrieved 7-11-06, from http://www. nytimes.com/2006/05/27/world/americas/27bears.html?ex=1306382 400&en=07809799811ff6cb&ei=5088&partner=rssnyt&emc=rss.

Kuhn, K. G., Campbell-Lendrum, D. H., Armstrong, B., & Davies, C. R. (2003). Malaria in Britain: Past, Present, and Future. *Proceedings of the National Academy of Sciences of the United States of America, 100*(17), 9997–10001.

Kundzewicz, Z. W., Graczyk, D., Maurer, T., Pinskwar, I., Radziejewski, M., Svensson, C., et al. (2005). Trend Detection in River Flow Series: 1. Annual Maximum Flow. *Hydrological Sciences Journal–Journal des Sciences Hydrologiques, 50*(5), 797–810.

Kunst, A. E., Looman, C. W. N., & Mackenbach, J. P. (1993). Outdoor Air-temperature and Mortality in the Netherlands—a Time-series Analysis. *American Journal of Epidemiology, 137*(3), 331–341.

Langford, I. H., & Bentham, G. (1995). The Potential Effects of Climate-change on Winter Mortality in England and Wales. *International Journal of Biometeorology, 38*(3), 141–147.

Larsen, J. (2003, October 9). Record Heat Wave in Europe Takes 35,000 Lives: Far Greater Losses May Lie Ahead. Earth Policy Institute. Retrieved 13-11-06, from http://www.earth-policy.org/Updates/Update29.htm.

Larson, L. A. (1994, July). Tough Lessons from Recent Floods—Special Section: America under Water. *USA Today Magazine.* Retrieved 23-12-06, from http://www.findarticles.com/p/articles/mi_m1272/is_n2590_v123/ai_15594504.

Latham, J. (1990). Control of Global Warming. *Nature, 347*(6291), 339–340.

Lau, K. M., Shen, S. S. P., Kim, K. M., & Wang, H. (2006). A Multimodel Study of the Twentieth-century Simulations of Sahel Drought from the 1970s to 1990s. *Journal of Geophysical Research–Atmospheres, 111*(D7).

Le Roy Ladurie, E. (1972). *Times of Feast, Times of Famine: A History of Climate since the Year 1000.* London: George Allen & Unwin.

Lean, G. (2004, May 2). Why Antarctica Will Soon Be the Only Place to Live. *The Independent.* Retrieved 12-11-06, from http://www.findarticles.com/p/articles/mi_qn4159/is_20040502/ai_n12755553.

Lean, G. (2005, January 23). Global Warming Approaching Point of No Return, Warns Leading Climate Expert. *The Independent.* Retrieved 26-1-07, from http://www.commondreams.org/headlines05/0123-01.htm.

Lebergott, S. (1993). *Pursuing Happiness: American Consumers in the Twentieth Century.* Princeton, N.J.: Princeton University Press.

Lebergott, S. (1995). Long-term Trends in the US Standard of Living. Pp. 149–160 in J. Simon (ed.), *State of Humanity.* Oxford: Blackwell.

Leggett, J. K. (2001). *The Carbon War: Global Warming and the End of the Oil Era.* New York: Routledge.

Lehmkuhl, F., & Owen, L. A. (2005). Late Quaternary Glaciation of Tibet and the Bordering Mountains: A Review. *Boreas, 34*(2), 87–100.

LibDem. (2006). Consultation Paper on Climate Change. UK Liberal Democrats. Retrieved 1-1-07, from http://consult.libdems.org.uk/climatechange/wp-content/uploads/2006/09/climate-change-cp84.pdf.

Lindzen, R. S. (2006, April 12). Climate of Fear. *The Wall Street Journal.* Retrieved 26-1-07, from http://www.opinionjournal.com/extra/?id=110008220.

Link, P. M., & Tol, R. S. J. (2004). Possible Economic Impacts of a Shutdown of the Thermohaline Circulation: An Application of FUND. *Portuguese Economic Journal, 3,* 99–114.

Lins, H. F., & Slack, J. R. (1999). Streamflow Trends in the United States. *Geophysical Research Letters, 26*(2), 227–230.

Lins, H. F., & Slack, J. R. (2005). Seasonal and Regional Characteristics of US Streamflow Trends in the United States from 1940 to 1999. *Physical Geography, 26*(6), 489–501.

Lomborg, B. (2001). *The Skeptical Environmentalist.* Cambridge: Cambridge University Press.

Lomborg, B. (ed.). (2004). *Global Crises, Global Solutions.* New York: Cambridge University Press.

Lomborg, B. (ed.). (2006). *How to Spend $50 Billion to Make the World a Better Place.* New York: Cambridge University Press.

Long, S. P., Ainsworth, E. A., Leakey, A. D. B., Nosberger, J., & Ort, D. R. (2006). Food for Thought: Lower-than-expected Crop Yield Stimulation with Rising CO_2 Concentrations. *Science, 312*(5782), 1918–1921.

Longstreth, J. (1999). Public Health Consequences of Global Climate Change in the United States—Some Regions May Suffer Disproportionately. *Environmental Health Perspectives, 107,* 169–179.

Lopez, A. D., Mathers, C. D., Ezzati, M., Jamison, D. T., & Murray, C. J. L. (2006). Global and Regional Burden of Disease and Risk Factors, 2001: Systematic Analysis of Population Health Data. *Lancet, 367*(9524), 1747–1757.

Lovell, J. (2006, November 28). Gaia Scientist Lovelock Predicts Planetary

Wipeout. Reuters. Retrieved 29-11-06, from http://www.alertnet.org/thenews/newsdesk/L28841108.htm.

Lovelock, J. E. (2006a, January 16). The Earth Is about to Catch a Morbid Fever That May Last as Long as 100,000 Years. *The Independent.* Retrieved 21-11-06, from http://comment.independent.co.uk/commentators/article338830.ece.

Lovelock, J. E. (2006b). *The Revenge of Gaia : Earth's Climate in Crisis and the Fate of Humanity.* New York: Basic Books.

Luthcke, S. B., Zwally, H. J., Abdalati, W., Rowlands, D. D., Ray, R. D., Nerem, R. S., et al. (2006). Recent Greenland Ice Mass Loss by Drainage System from Satellite Gravity Observations. *Science, 314*(5803), 1286–1289.

Lynas, M. (2004). *High Tide: The Truth about Our Climate Crisis.* New York: Picador.

Lynas, M. (2006, May 19). Climate Denial Ads to Air on US National Television. Retrieved 26-1-07, from http://www.marklynas.org/2006/5/19/climate-denial-ads-to-air-on-us-national-television.

Madden, A. H. (1945). A Brief History of Medical Entomology in Florida. *The Florida Entomologist, 28*(1), 1–7.

Maddison, A. (2006). World Population, GDP and Per Capita GDP, 1–2003 AD. Retrieved 8-12-06, from http://www.ggdc.net/Maddison/.

Mahlman, J. D. (1997). Uncertainties in Projections of Human-caused Climate Warming. *Science, 278*(5342), 1416–1417.

Manne, A., & Richels, R. (2004). US Rejection of the Kyoto Protocol: The Impact on Compliance Costs and CO_2 Emissions. *Energy Policy, 32*(4), 447–454.

Marland, G., Andres, B., & Boden, T. (2006). Global, Regional, and National CO_2 Emissions. In *Trends: A Compendium of Data on Global Change.* Carbon Dioxide Information Analysis Center.

Marshall, G. J., Orr, A., van Lipzig, N. P. M., & King, J. C. (2006). The Impact of a Changing Southern Hemisphere Annular Mode on Antarctic Peninsula Summer Temperatures. *Journal of Climate, 19*(20), 5388–5404.

Martens, P., Kovats, R. S., Nijhof, S., de Vries, P., Livermore, M. T. J., Bradley, D. J., et al. (1999). Climate Change and Future Populations at Risk of Malaria. *Global Environmental Change, 9*(suppl. 1), S89–S107.

Martens, W. J. M. (1998). Climate Change, Thermal Stress and Mortality Changes. *Social Science and Medicine, 46*(3), 331–344.

Matthews, J. A., Berrisford, M. S., Dresser, P. Q., Nesje, A., Dahl, S. O., Bjune, A. E., et al. (2005). Holocene Glacier History of Bjornbreen and Climatic Reconstruction in Central Jotunheimen, Norway, Based on

Proximal Glaciofluvial Stream-bank Mires. *Quaternary Science Reviews, 24*(1–2), 67–90.

Matthews, J. A., & Briffa, K. R. (2005). The "Little Ice Age": Re-evaluation of an Evolving Concept. *Geografiska Annaler Series A–Physical Geography, 87A*(1), 17–36.

Matthews, N. (2000). The Attack of the Killer Architects. *Travel Holiday, 183*(7), 80–88.

Maugeri, M., Buffoni, L., Delmonte, B., & Fassina, A. (2002). Daily Milan Temperature and Pressure Series (1763–1998): Completing and Homogenising the Data. *Climatic Change, 53*(1–3), 119–149.

McCallum, E., & Heming, J. (2006). Hurricane Katrina: An Environmental Perspective. *Philosophical Transactions of the Royal Society A: Mathematical, Physical and Engineering Sciences, 364*(1845), 2099–2115.

McCarthy, M. (2005, February 3). Global Warming: Scientists Reveal Timetable. *The Independent.* Retrieved 3-1-07, from http://www.commondreams.org/headlines05/0203-04.htm.

McCarthy, M. (2006, February 11). Global Warming: Passing the "Tipping Point." *The Independent.* Retrieved 6-11-06, from http://www.countercurrents.org/cc-mccarthy110206.htm.

McKibben, B. (2004, September–October). The Submerging World. *Orion.*

McMichael, A. J., Campbell-Lendrum, D. H., Corvalán, C. F., Ebi, K. L., Githeko, A. K., Scheraga, J. D., et al. (eds.). (2003). *Climate Change and Human Health.* Geneva: World Health Organization. See http://www.who.int/globalchange/publications/cchhbook/en/index.html.

McMichael, A. J., Woodruff, R. E., & Hales, S. (2006). Climate Change and Human Health: Present and Future Risks. *Lancet, 367*(9513), 859–869.

Mégroz, R. L. (1937). The World-wide Scourge of Malaria. *Contemporary Review, 151,* 349–356.

Meissner, K. J., & Clark, P. U. (2006). Impact of Floods versus Routing Events on the Thermohaline Circulation. *Geophysical Research Letters, 33*(15).

Mendelsohn, R. (2004). Perspective Paper 1.1 on Climate Change. Pp. 44–48 in Lomborg, 2004.

Mendelsohn, R. (2007). A Critique of the Stern Report. *Regulation,* Winter, 42–46.

Merali, Z. (2006, November 7). No New Ice Age for Western Europe. *New Scientist,* 13.

Metcalf, G. E., & Hassertt, K. A. (1997). Measuring the Energy Savings from Home Improvement Investments: Evidence from Monthly Billing Data. National Bureau of Economic Research Working Paper 6074. Cambridge: National Bureau of Economic Research.

226 *Literature*

Michaels, P. J. (2004, November 22). Polar Disasters: More Predictable Distortions of Science. Retrieved 7-11-06, from http://www.cato.org/pub_display.php?pub_id=2888.

Michaels, P. J., Knappenberger, P. C., Balling, R. C., & Davis, R. E. (2000). Observed Warming in Cold Anticyclones. *Climate Research, 14*(1), 1–6.

Miguel, E. (2005). Poverty and Witch Killing. *Review of Economic Studies, 72*(4), 1153–1172.

Milliken, M. (2004, December 10). "After Kyoto" Takes Center Stage at Climate Talks. Reuters. Retrieved 18-11-06, from http://www.climateark.org/shared/reader/welcome.aspx?linkid=37207.

Mills, A., & Shillcutt, S. (2004). Communicable Diseases. Pp. 62–114 in Lomborg, 2004.

Mills, E., & Lecomte, E. (2006, August). From Risk to Opportunity: How Insurers Can Proactively and Profitably Manage Climate Change. *Ceres.* Retrieved 21-12-06, from http://www.ceres.org/pub/docs/Ceres_Insurance_Climate_%20Report_082206.pdf.

Milly, P. C. D., Wetherald, R. T., Dunne, K. A., & Delworth, T. L. (2002). Increasing Risk of Great Floods in a Changing Climate. *Nature, 415*(6871), 514–517.

Mitchell, J. K. (2003). European River Floods in a Changing World. *Risk Analysis, 23*(3), 567–574.

Moberg, A., & Bergstrom, H. (1997). Homogenization of Swedish Temperature Data: 3. The Long Temperature Records from Uppsala and Stockholm. *International Journal of Climatology, 17*(7), 667–699.

Moberg, A., Sonechkin, D. M., Holmgren, K., Datsenko, N. M., & Karlen, W. (2005). Highly Variable Northern Hemisphere Temperatures Reconstructed from Low- and High-resolution Proxy Data. *Nature, 433*(7026), 613–617.

Monaghan, A. J., & Bromwich, D. H. (2006). A High Spatial Resolution Record of Near-surface Temperature over WAIS during the Past 5 Decades. Thirteenth Annual WAIS Workshop. Retrieved 14-12-06, from http://igloo.gsfc.nasa.gov/wais/pastmeetings/Sched06.htm.

Monbiot, G. (2006). *Heat: How to Stop the Planet Burning.* London: Allen Lane.

Monnett, C., Gleason, J. S., & Rotterman, L. M. (2005, December). Potential Effects of Diminished Sea Ice on Open-water Swimming, Mortality, and Distribution of Polar Bears during Fall in the Alaskan Beaufort Sea. Retrieved 7-11-06, from http://www.mms.gov/alaska/ess/Poster%20Presentations/MarineMammalConference-Dec2005.pdf.

Morris, E. M., & Mulvaney, R. (2004). Recent Variations in Surface Mass Balance of the Antarctic Peninsula Ice Sheet. *Journal of Glaciology, 50*(169), 257–267.

Mudelsee, M., Borngen, M., Tetzlaff, G., & Grunewald, U. (2003). No Upward Trends in the Occurrence of Extreme Floods in Central Europe. *Nature, 425*(6954), 166–169.

Mudelsee, M., Deutsch, M., Borngen, M., & Tetzlaff, G. (2006). Trends in Flood Risk of the River Werra (Germany) over the Past 500 Years. *Hydrological Sciences Journal–Journal des Sciences Hydrologiques, 51*(5), 818–833.

Murray, T. (2006). Climate Change: Greenland's Ice on the Scales. *Nature, 443*(7109), 277–278.

Nakicenovic, N., & IPCC WG III (2000). *Special Report on Emissions Scenarios: A Special Report of Working Group III of the Intergovernmental Panel on Climate Change.* New York: Cambridge University Press. See http://www.grida.no/climate/ipcc/emission/index.htm.

NCEP. (2006). Global Surface Air Temperature, Annual Average 1982–94. National Center for Environmental Prediction.

Nemet, G. F., & Kammen, D. M. (2007). US Energy Research and Development: Declining Investment, Increasing Need, and the Feasibility of Expansion. *Energy Policy, 35*(1), 746–755.

NERI. (1998). *Natur og Miljø 1997: Påvirkninger og Tilstand* [Nature and environment 1997: Effects and state]. National Environmental Research Institute of Denmark.

New Scientist Anon. (2005, March 26). The Edge of the Abyss. *New Scientist,* 5.

Nicholls, R. J. (2004). Coastal Flooding and Wetland Loss in the 21st Century: Changes under the SRES Climate and Socio-economic Scenarios. *Global Environmental Change—Human and Policy Dimensions, 14*(1), 69–86.

Nicholls, R. J., & Tol, R. S. J. (2006). Impacts and Responses to Sea-level Rise: A Global Analysis of the SRES Scenarios over the Twenty-first Century. *Philosophical Transactions of the Royal Society A: Mathematical, Physical and Engineering Sciences, 364*(1841), 1073–1095.

NOAA. (2006). Hurricane History. National Hurricane Center. Retrieved 20-12-06, from http://www.nhc.noaa.gov/HAW2/english/history.shtml.

Nohara, D., Kitoh, A., Hosaka, M., & Oki, T. (2006). Impact of Climate Change on River Discharge Projected by Multimodel Ensemble. *Journal of Hydrometeorology, 7*(5), 1076–1089.

Nordhaus, W. D. (1992). An Optimal Transition Path for Controlling Greenhouse Gases. *Science, 258*(5086), 1315–1319.

Nordhaus, W. D. (1994). *Managing the Global Commons: The Economics of Climate Change.* Cambridge: MIT Press.

Nordhaus, W. D. (2001). Climate Change—Global Warming Economics. *Science, 294*(5545), 1283–1284.

Nordhaus, W. D. (2006a). After Kyoto: Alternative Mechanisms to Control Global Warming. *American Economic Review, 96*(2), 31–34.

Nordhaus, W. D. (2006b). DICE Model. Retrieved 22-11-06, from http://nordhaus.econ.yale.edu/dicemodels.htm.

Nordhaus, W. D. (2006c). Life after Kyoto: Alternative Approaches to Global Warming Policies. Prepared for the Annual Meetings of the American Economic Association, Boston, Massachusetts, January 5–8, 2006. Retrieved 18-11-06, from http://nordhaus.econ.yale.edu/kyoto_long_2005.pdf.

Nordhaus, W. D. (2006d). RICE Model. Retrieved 27-11-06, from http://www.econ.yale.edu/~nordhaus/homepage/dice_section_vi.html.

Nordhaus, W. D. (2006e). The *Stern Review* on the Economics of Climate Change. Retrieved 24-1-07, from http://nordhaus.econ.yale.edu/SternReviewD2.pdf.

Nordhaus, W. D. (ed.). (1998). *Economics and Policy Issues in Climate Change.* Washington, D.C.: Resources for the Future.

Nordhaus, W. D., & Boyer, J. (2000). *Warming the World: Economic Models of Global Warming.* Cambridge: MIT Press. See http://www.econ.yale.edu/~nordhaus/homepage/web%20table%20of%20contents%20102599.htm.

Nordhaus, W. D., & Yang, Z. L. (1996). A Regional Dynamic General-equilibrium Model of Alternative Climate-change Strategies. *American Economic Review, 86*(4), 741–765.

Norris, S., Rosentrater, L., & Eid, P. M. (2002). Polar Bears at Risk: A WWF Status Report. Gland, Switzerland: WWF-World Wide Fund for Nature. Retrieved 6-11-06, from http://www.wwf.org.uk/filelibrary/pdf/polar_bears_at_risk_report.pdf.

NRDC. (2006). Global Warming Basics: What It Is, How It's Caused, and What Needs to Be Done to Stop It. National Resources Defense Council. Retrieved 17-12-06, from http://www.nrdc.org/globalWarming/f101.asp.

O'Neill, B. (2006a, October 6). Global Warming: The Chilling Effect on Free Speech. Spiked. Retrieved 26-1-07, from http://www.spiked-online.com/index.php?/site/article/1782/.

O'Neill, B. (2006b, November 7). A March of Middle-class Miserabilists. Spiked. Retrieved 27-1-07, from http://www.spiked-online.com/index.php?/site/article/2071.

O'Neill, M. S., Hajat, S., Zanobetti, A., Ramirez-Aguiler, M., & Schwartz, J. (2005). Impact of Control for Air Pollution and Respiratory Epidemics on the Estimated Associations of Temperature and Daily Mortality. International Journal of Biometeorology, 50(2), 121–129.

O'Rangers, E. A. (2005, January). NASA Spin-offs: Bringing Space Down to Earth. *AdAstra: The Magazine of the National Space Society*. Retrieved 19-1-07, from http://www.space.com/adastra/adastra_spinoffs_050127.html.

OECD. (2005). *OECD in Figures*. OECD Observer.

OECD. (2006). OECD Factbook 2006. Paris: Organization for Economic Cooperation and Development. Retrieved 10-11-06, from http://titania.sourceoecd.org/vl=4901411/cl=16/nw=1/rpsv/fact2006/.

Oerlemans, J. (2000). Holocene Glacier Fluctuations: Is the Current Rate of Retreat Exceptional? *Annals of Glaciology, 31*, 39–44.

Oerlemans, J. (2005). Extracting a Climate Signal from 169 Glacier Records. *Science, 308*(5722), 675–677.

Oerlemans, J., Bassford, R. P., Chapman, W., Dowdeswell, J. A., Glazovsky, A. F., Hagen, J. O., et al. (2005). Estimating the Contribution of Arctic Glaciers to Sea-level Change in the Next 100 Years. *Annals of Glaciology, 42*, 230–236.

Oster, E. (2004). Witchcraft, Weather and Economic Growth in Renaissance Europe. *Journal of Economic Perspectives, 18*(1), 215–228.

Owen, J. (2005, November 30). "Mini Ice Age" May Be Coming Soon, Sea Study Warns. National Geographic News. Retrieved 27-12-06, from http://news.nationalgeographic.com/news/2005/11/1130_051130_ice_age.html.

Parizek, B. R., & Alley, R. B. (2004). Implications of Increased Greenland Surface Melt under Global-warming Scenarios: Ice-sheet Simulations. *Quaternary Science Reviews, 23*(9–10), 1013–1027.

Parkinson, C. L. (2006). Earth's Cryosphere: Current State and Recent Changes. *Annual Review of Environment and Resources, 31*(1), 33–60.

Parry, M. (2004). Global Impacts of Climate Change under the SRES Scenarios. *Global Environmental Change, 14*(1), 1–1.

Parry, M., Rosenzweig, C., & Livermore, M. (2005). Climate Change, Global Food Supply and Risk of Hunger. *Philosophical Transactions of the Royal Society B: Biological Sciences, 360*(1463), 2125–2138.

Parry, M. L., Rosenzweig, C., Iglesias, A., Livermore, M., & Fischer, G. (2004). Effects of Climate Change on Global Food Production under SRES Emissions and Socio-economic Scenarios. *Global Environmental Change, 14*(1), 53–67.

Pascual, M., Ahumada, J. A., Chaves, L. F., Rodo, X., & Bouma, M. (2006). Malaria Resurgence in the East African Highlands: Temperature Trends Revisited. *Proceedings of the National Academy of Sciences of the United States of America, 103*(15), 5829–5834.

Patz, J. A., Campbell-Lendrum, D., Holloway, T., & Foley, J. A. (2005).

Impact of Regional Climate Change on Human Health. *Nature,* *438*(7066), 310–317.

Patz, J. A., Hulme, M., Rosenzweig, C., Mitchell, T. D., Goldberg, R. A., Githeko, A. K., et al. (2002). Climate Change—Regional Warming and Malaria Resurgence. *Nature, 420*(6916), 627–628.

Pearce, D. (2003). The Social Cost of Carbon and Its Policy Implications. *Oxford Review of Economic Policy, 19*(3), 362–384.

Pearce, F. (2001). We Are All Guilty! It's Official, People Are to Blame for Global Warming. *New Scientist, 169*(2275), 5.

Pearce, F. (2005a, December 3). Faltering Currents Trigger Freeze Fear. *New Scientist,* 6.

Pearce, F. (2005b, August 27). Global Warming: The Flaw in the Thaw. *New Scientist,* 26.

Pearce, F. (2005c, December 24). Review 2005: Climate Going Crazy. *New Scientist,* 16.

Pearce, F. (2006). *The Last Generation: How Nature Will Take Her Revenge for Climate Change.* London: Eden Project Books.

Petrow, T., Thieken, A. H., Kreibich, H., Bahlburg, C. H., & Merz, B. (2006). Improvements on Flood Alleviation in Germany: Lessons Learned from the Elbe Flood in August 2002. *Environmental Management, 38*(5), 717–732.

Pew Center on Global Climate Change. (2006). *Regional Initiatives.* Retrieved 6-11-06, from http://www.pewclimate.org/what_s_being _done/in_the_states/regional_initiatives.cfm?preview=1.

Pfister, C., Weingartner, R., & Luterbacher, J. (2006). Hydrological Winter Droughts over the Last 450 Years in the Upper Rhine Basin: A Methodological Approach. *Hydrological Sciences Journal–Journal des Sciences Hydrologiques, 51*(5), 966–985.

Pielke, R. A., Jr., (1999). Nine Fallacies of Floods. *Climatic Change, 42*(2), 413–438.

Pielke, R. A., Jr. (2005). Misdefining "Climate Change": Consequences for Science and Action. *Environmental Science and Policy, 8*(6), 548–561.

Pielke, R. A., Jr., & Downton, M. W. (2000). Precipitation and Damaging Floods: Trends in the United States, 1932–97. *Journal of Climate, 13*(20), 3625–3637.

Pielke, R. A., Jr., & Landsea, C. W. (1998). Normalized Hurricane Damages in the United States: 1925–95. *Weather and Forecasting, 13*(3), 621–631.

Pielke, R. A., Jr. (2006). Disasters, Death, and Destruction: Making Sense of Recent Calamities. *Oceanography, 19*(2), 138–147.

Pielke, R. A., Jr., Gratz, J., Landsea, C. W., Collins, D., Saunders, M. A., &

Musulin, R. (2007). Normalized Hurricane Damages in the United States: 1900–2005. *Natural Hazards Review* (submitted). Retrieved 19-12-06, from http://sciencepolicy.colorado.edu/publications/special/normalized_hurricane_damages.html.

Pielke, R. A., Jr., Klein, R., & Sarewitz, D. (2000). Turning the Big Knob: An Evaluation of the Use of Energy Policy to Modulate Future Climate Impacts. *Energy and Environment, 11,* 255–276. Retrieved 20-12-06, from http://sciencepolicy.colorado.edu/about_us/meet_us/roger_pielke/knob/text.html.

Pinter, N. (2005). Environment—One Step Forward, Two Steps Back on US Floodplains. *Science, 308*(5719), 207–208.

Pinter, N., & Heine, R. A. (2005). Hydrodynamic and Morphodynamic Response to River Engineering Documented by Fixed-discharge Analysis, Lower Missouri River, USA. *Journal of Hydrology, 302*(1–4), 70–91.

Plumb, C. (2003, December 11). Climate Change Death Toll Put at 150,000. Reuters. Retrieved 1-1-07, from http://www.commondreams.org/headlines03/1211-13.htm.

Plummer, N., Salinger, M. J., Nicholls, N., Suppiah, R., Hennessy, K. J., Leighton, R. M., et al. (1999). Changes in Climate Extremes over the Australian Region and New Zealand during the Twentieth Century. *Climatic Change, 42*(1), 183–202.

Postman, A. (2006, October 5). The Energy Diet. *The New York Times.* Retrieved 21-11-06, from http://www.stopglobalwarming.org/sgw_read.asp?id=1128441052006.

Prodi, R. (2004, July 15). Climate Change—the Real Threat to Global Peace. Retrieved 6-11-06, from http://www.europa-eu-un.org/articles/en/article_3678_en.htm.

Przybylak, R. (2000). Temporal and Spatial Variation of Surface Air Temperature over the Period of Instrumental Observations in the Arctic. *International Journal of Climatology, 20*(6), 587–614.

Pudsey, C. J., Murray, J. W., Appleby, P., & Evans, J. (2006). Lee Shelf History from Petrographic and Foraminiferal Evidence, Northeast Antarctic Peninsula. *Quaternary Science Reviews, 25*(17–18), 2357–2379.

Pullella, P. (2005, May 27). Global Warming Will Increase World Hunger. Reuters. Retrieved 3-1-07, from http://www.globalpolicy.org/socecon/envronmt/2005/0527warming.htm.

Purcell, K. (2006). Gates Foundation Invests $42.6 Million in Malaria Drug Research. *HerbalGram: The Journal of the American Botanical Council, 69*(24), 252. Retrieved 30-12-06, from http://www.herbalgram.org/herbalgram/articleview.asp?a=2919&p=Y.

Reiter, P. (2000). From Shakespeare to Defoe: Malaria in England in the Little Ice Age. *Emerging Infectious Diseases, 6*(1), 1–10. Retrieved 7-12-06, from http://www.cdc.gov/ncidod/eid/vol6no1/reiter.htm.

Reiter, P., Thomas, C. J., Atkinson, P. M., Hay, S. I., Randolph, S. E., Rogers, D. J., et al. (2004). Global Warming and Malaria: A Call for Accuracy. *Lancet Infectious Diseases, 4*(6), 323–324.

Reuters. (2001, November 7). African Mountains Snow Melting Down—Greenpeace. Retrieved 7-12-06, from http://www.planetark.org/avantgo/dailynewsstory.cfm?newsid=13154.

Reuters. (2002, September 3). Sound of Conflict Blurs Earth Summit Rhetoric. Retrieved 22-12-06, from http://www.planetark.org/avantgo/dailynewsstory.cfm?newsid=17557.

Revenga, C., Brunner, J., Henninger, N., Payne, R., & Kassem, K. (2000). *Pilot Analysis of Global Ecosystems: Freshwater Systems.* World Resources Institute. Retrieved 7-1-07, from http://www.wri.org/biodiv/pubs_description.cfm?pid=3056.

Richey, L. A. (2003). HIV/AIDS in the Shadows of Reproductive Health Interventions. *Reproductive Health Matters, 11*(22), 30–35.

Rijsberman, F. (2004). Sanitation and Access to Clean Water. Pp. 498–527 in Lomborg, 2004.

Rising Tide. (2007). Hall of Shame. Retrieved 26-1-07, from http://risingtide.org.uk/pages/voices/hall_shame.htm.

Roberts, D. (2006a, May 9). Al Revere: An Interview with Accidental Movie Star Al Gore. Grist. Retrieved 26-1-07, from http://www.grist.org/news/maindish/2006/05/09/roberts/index.html.

Roberts, D. (2006b, September 19). The Denial Industry. Grist. Retrieved 26-1-07, from http://gristmill.grist.org/print/2006/9/19/11408/1106.

Robinson, J. P., & Godbey, G. (1997). *Time for Life: The Surprising Ways Americans Use Their Time.* University Park: Pennsylvania State University Press.

Rogers, D. J., & Randolph, S. E. (2000). The Global Spread of Malaria in a Future, Warmer World. *Science, 289*(5485), 1763–1766.

Rolling Stone. (2007). Extreme Makeover: Images of Planetwide Damage Caused by Global Warming, with Selected Quotes from Our Interview with Al Gore and Ten Ways You Can Help. *Rolling Stone.* Retrieved 2-1-07, from http://www.rollingstone.com/politics/story/10698217/extreme_makeover.

Rosenberg, T. (2004, April 11). What the World Needs Now Is DDT. *The New York Times.* Retrieved 30-12-06, from http://query.nytimes.com/gst/fullpage.html?res=9F0DEEDA1738F932A25757C0A9629C8B63&sec=health&spon=&pagewanted=print.

Rosenfeld, A. H., Akbari, H., Romm, J. J., & Pomerantz, M. (1998). Cool Communities: Strategies for Heat Island Mitigation and Smog Reduction. *Energy and Buildings, 28*(1), 51–62.

Rosenzweig, C., & Parry, M. L. (1994). Potential Impact of Climate-change on World Food-supply. *Nature, 367*(6459), 133–138.

Rosenzweig, C., Solecki, W., Parshall, L., Gaffin, S., Lynn, B., Goldberg, R., et al. (2006, January 31). Mitigating New York City's Heat Island with Urban Forestry, Living Roofs, and Light Surfaces. Paper presented at the Sixth Symposium on the Urban Environment, AMS Forum: Managing Our Physical and Natural Resources: Successes and Challenges. Retrieved 17-11-06, from http://ams.confex.com/ams/Annual2006/techprogram/paper_103341.htm.

Rosing-Asvid, A. (2006). The Influence of Climate Variability on Polar Bear (Ursus Maritimus) and Ringed Seal (Pusa Hispida) Population Dynamics. *Canadian Journal of Zoology, 84*, 357–364. See http://article.pubs.nrc-cnrc.gc.ca/ppv/RPViewDoc?_handler_=HandleInitialGet&journal=cjz&volume=84&calyLang=eng&articleFile=z06-001.pdf.

Roy Morgan Research. (2006, November 2). Protecting the Environment More Important Than the War on Terror. Roy Morgan International. Retrieved 30-11-06, from http://www.roymorgan.com/news/polls/2006/4100/.

Ruhland, K., Phadtare, N. R., Pant, R. K., Sangode, S. J., & Smol, J. P. (2006). Accelerated Melting of Himalayan Snow and Ice Triggers Pronounced Changes in a Valley Peatland from Northern India. *Geophysical Research Letters, 33*(15).

Runci, P. (2005). Energy R&D Investment Patterns in IEA Countries: An Update. Pacific Northwest National Laboratory/Joint Global Change Research Institute Technical Paper PNWD-3581. Retrieved 19-1-07, from http://www.globalchange.umd.edu/?energytrends&page=lea.

Saaroni, H., Ben-Dor, E., Bitan, A., & Potchter, O. (2000). Spatial Distribution and Microscale Characteristics of the Urban Heat Island in Tel-Aviv, Israel. *Landscape and Urban Planning, 48*(1–2), 1–18.

Sanchez, P., Swaminathan, M. S., Dobie, P., & Yuksel, N. (2005). *Halving Hunger: It Can Be Done*. UN Millenium Project Task Force on Hunger. Retrieved 4-1-07, from http://www.unmillenniumproject.org/reports/tf_hunger.htm.

Sarewitz, D., & Pielke, R. A., Jr. (2005, January 17). Rising Tide. *The New Republic*, 10.

Sarewitz, D., & Pielke, R. A., Jr. (2007). The Steps Not Yet Taken. In Kleinman, D. L., Cloud-Hansen, K., Matta C., & Handelsman, J. (eds.), *Controversies in Science and Technology, Volume 2*. Retrieved 17-1-07, from

http://sciencepolicy.colorado.edu/prometheus/archives/climate
_change/001048the_steps_not_yet_ta.html.

Schäfer, A. (2006). Long-term Trends in Global Passenger Mobility. *The Bridge, 36*(4), 24–32. Retrieved 30-1-07, from http://www.nae.edu/nae/bridgecom.nsf/weblinks/MKEZ-6WHQ3Q?OpenDocument.

Schapira, A. (2006). DDT: A Polluted Debate in Malaria Control. *Lancet, 368*(9553), 2111–2113.

Schell, J. (1989, October). *Discover,* 45–48.

Schiermeier, Q. (2006). A Sea Change. *Nature, 439*(7074), 256–260.

Schipper, L. J., Haas, R., & Sheinbaum, C. (1996). Recent Trends in Residential Energy Use in OECD Countries and Their Impact on Carbon Dioxide Emissions: A Comparative Analysis of the Period 1973–1992. *Mitigation and Adaptation Strategies for Global Change, 1*(2), 167–196.

Schneeberger, C., Blatter, H., Abe-Ouchi, A., & Wild, M. (2003). Modelling Changes in the Mass Balance of Glaciers of the Northern Hemisphere for a Transient 2 x CO_2 Scenario. *Journal of Hydrology, 282*(1–4), 145–163.

Schwartz, P., & Randall, D. (2003, October). An Abrupt Climate Change Scenario and Its Implications for United States National Security. Commissioned by the Pentagon. Retrieved 25-12-06, from http://www.grist.org/pdf/AbruptClimateChange2003.pdf.

Schwoon, M., & Tol, R. S. J. (2006). Optimal CO_2-abatement with Socio-economic Inertia and Induced Technological Change. *Energy Journal, 27*(4), 25–59.

Seager, R. (2006). The Source of Europe's Mild Climate. *American Scientist, 94*(4), 334–341.

SEHN. (2007). Precautionary Principle: FAQs. Science and Environmental Health Network. Retrieved 30-1-07, from http://www.sehn.org/ppfaqs.html.

Serageldin, I. (1995). Toward Sustainable Management of Water Resources. *Directions in Development, 14,* 910.

Shanks, G. D. (2006). Treatment of Falciparum Malaria in the Age of Drug-resistance. *Journal of Postgraduate Medicine, 52*(4), 277–280.

Shanks, G. D., Hay, S. I., Omumbo, J. A., & Snow, R. W. (2005). Malaria in Kenya's Western Highlands. *Emerging Infectious Diseases, 11*(9), 1425–1432. See http://www.cdc.gov/ncidod/EID/vol11no09/04-1131.htm.

Shanks, G. D., Hay, S. I., Stern, D. I., Biomndo, K., & Snow, R. W. (2002). Meteorologic Influences on Plasmodium Falciparum Malaria in the Highland Tea Estates of Kericho, Western Kenya. *Emerging Infectious Diseases, 8*(12), 1404–1408.

Shepherd, A., & Wingham, D. (2007). Recent Sea-level Contributions of the Antarctic and Greenland Ice Sheets. *Science, 315*(5818), 1529–1532.

Shute, N., Hayden, T., Petit, C. W., Sobel, R. K., Whitelaw, K., & Whitman, D. (2001, February 5). The Weather Turns Wild. *U.S. News & World Report,* 44–50.

Sierra Club. (2007). Smart Energy Solutions. Retrieved 29-1-07, from http://www.sierraclub.org/energy/.

Simms, A., Magrath, J., & Reid, H. (2004). *Up in Smoke.* London: New Economics Foundation with the Working Group on Climate Change. Retrieved 3-1-07, from http://www.neweconomics.org/gen/uploads/igeebque0l3nvy455whn42vs19102004202736.pdf.

Simonovic, S. P. (2002). World Water Dynamics: Global Modeling of Water Resources. *Journal of Environmental Management, 66*(3), 249–267.

Singh, P., Arora, M., & Goel, N. K. (2006). Effect of Climate Change on Runoff of a Glacierized Himalayan Basin. *Hydrological Processes, 20*(9), 1979–1992.

Singh, P., & Bengtsson, L. (2004). Hydrological Sensitivity of a Large Himalayan Basin to Climate Change. *Hydrological Processes, 18*(13), 2363–2385.

Singh, P., & Bengtsson, L. (2005). Impact of Warmer Climate on Melt and Evaporation for the Rainfed, Snowfed and Glacierfed Basins in the Himalayan Region. *Journal of Hydrology, 300*(1–4), 140–154.

Small, D., Islam, S., & Vogel, R. M. (2006). Trends in Precipitation and Streamflow in the Eastern US: Paradox or Perception? *Geophysical Research Letters, 33*(3).

Smith, D. (2005, December 1). Scientists Forecast Global Cold Snap. *Sydney Morning Herald.* Retrieved 27-12-06, from http://www.smh.com.au/news/science/scientists-forecast-global-cold-snap/2005/12/01/1133311132663.html.

Snow, R., Ikoku, A., Omumbo, J., & Ouma, J. (1999). *The Epidemiology, Politics and Control of Malaria Epidemics in Kenya: 1900–1998.* Report prepared for Roll Back Malaria, Resource Network on Epidemics, World Health Organization. Retrieved 28-12-06, from http://www.who.int/malaria/docs/ek_report_toc1.htm#toc.

Snow, R. W., Guerra, C. A., Noor, A. M., Myint, H. Y., & Hay, S. I. (2005). The Global Distribution of Clinical Episodes of Plasmodium Falciparum Malaria. *Nature, 434*(7030), 214–217.

Snow, R. W., & Omumbo, J. A. (2006). Malaria. Pp. 195–213 in Jamison et al., 2006.

Soini, E. (2005). Land Use Change Patterns and Livelihood Dynamics on the Slopes of Mt. Kilimanjaro, Tanzania. *Agricultural Systems, 85*(3), 306–323.

Stern, N. (2006). *Stern Review on the Economics of Climate Change.* Her Majesty's Treasury, United Kingdom. Retrieved 24-11-06, from http://www.hm-treasury.gov.uk/independent_reviews/stern_review_economics_climate_change/stern_review_report.cfm.

Stipp, D. (2004, February 9). The Pentagon's Weather Nightmare. *Fortune*. Retrieved 25-12-06, from http://money.cnn.com/magazines/fortune/fortune_archive/2004/02/09/360120/index.htm.

Stirling, I., Lunn, N. J., & Iacozza, J. (1999). Long-term Trends in the Population Ecology of Polar Bears in Western Hudson Bay in Relation to Climatic Change. *Arctic, 52*(3), 294–306.

Stouffer, R. J., Yin, J., Gregory, J. M., Dixon, K. W., Spelman, M. J., Hurlin, W., et al. (2006). Investigating the Causes of the Response of the Thermohaline Circulation to Past and Future Climate Changes. *Journal of Climate, 19*(8), 1365–1387.

Streutker, D. R. (2003). Satellite-measured Growth of the Urban Heat Island of Houston, Texas. *Remote Sensing of Environment, 85*(3), 282–289.

Svensson, C., Kundzewicz, Z. W., & Maurer, T. (2005). Trend Detection in River Flow Series: 2. Flood and Low-flow Index Series. *Hydrological Sciences Journal–Journal des Sciences Hydrologiques, 50*(5), 811–824. doi: 10.1623/hysj.2005.50.5.811.

Swellengrebel, N. H. (1950). The Malaria Epidemic of 1943–1946 in the Province of North-Holland. *Transactions of the Royal Society of Tropical Medicine and Hygiene, 43*(5), 445–461.

Swiss Re. (1999). Natural Catastrophes and Man-made Disasters 1998: Storms, Hail and Ice Cause Billion-dollar Losses. Swiss Reinsurance Company.

Synnefa, A., Santamouris, M., & Livada, I. (2006). A Study of the Thermal Performance of Reflective Coatings for the Urban Environment. *Solar Energy, 80*(8), 968–981.

Taylor, M. (2006, May 1). Silly to Predict Their Demise: Startling Conclusion to Say They Will Disappear within 25 Years and Surprise to Many Researchers. *Toronto Star*.

Teklehaimanot, A., et al. & UN Millennium Project Working Group on Malaria. (2005). *Coming to Grips with Malaria in the New Millennium*. London: Earthscan. Retrieved 17-3-07, from http://www.unmillenniumproject.org/documents/malaria-complete-lowres.pdf.

Tereshchenko, I. E., & Filonov, A. E. (2001). Air Temperature Fluctuations in Guadalajara, Mexico, from 1926 to 1994 in Relation to Urban Growth. *International Journal of Climatology, 21*(4), 483–494.

Thijssen, J. (2001). Mount Kilimanjaro Expedition. Greenpeace. Retrieved 7-12-06, from http://archive.greenpeace.org/climate/climatecountdown/kilimanjaro.htm.

Thompson, K. (1969). Irrigation as a Menace to Health in California: A Nineteenth Century View. *Geographical Review, 59*(2), 195–214.

Thorndycraft, V. R., Barriendos, M., Benito, G., Rico, M., & Casas, A. (2006). The Catastrophic Floods of AD 1617 in Catalonia (Northeast Spain) and Their Climatic Context. *Hydrological Sciences Journal–Journal des Sciences Hydrologiques, 51*(5), 899–912.

Time. (2006, April 3). Be Worried. Be *Very* Worried. Retrieved 6-11-06, from http://www.time.com/time/covers/0,16641,20060403,00.html.

Timmons, H. (2006, October 30). U.K. Fears Disaster in Climate Change. *International Herald Tribune.* Retrieved 24-1-07, from http://www.iht.com/bin/print.php?id=3334967.

Tindale, S. (2005). Two-thirds of Energy Wasted by Antiquated System. Greenpeace.

Tol, R. S. J. (2002a). Estimates of the Damage Costs of Climate Change: Part 1: Benchmark Estimates. *Environmental and Resource Economics, 21*(1), 47–73.

Tol, R. S. J. (2002b). Estimates of the Damage Costs of Climate Change: Part 2: Dynamic Estimates. *Environmental and Resource Economics, 21*(2), 135–160.

Tol, R. S. J. (2004). The Double Trade-off between Adaptation and Mitigation for Sea Level Rise: An Application of FUND (vol. FNU-48). Hamburg University and Centre for Marine and Atmospheric Science. Retrieved 17-12-06, from http://www.uni-hamburg.de/Wiss/FB/15/Sustainability/slradaptmitigatewp.pdf.

Tol, R. S. J. (2005). The Marginal Damage Costs of Carbon Dioxide Emissions: An Assessment of the Uncertainties. *Energy Policy, 33*(16), 2064–2074.

Tol, R. S. J. (2006). The Stern Review of the Economics of Climate Change: A Comment. *Energy and Environment, 17*(6), 977–981.

Tol, R. S. J. (2007). Europe's Long-term Climate Target: A Critical Evaluation. *Energy Policy, 35*(1), 424–432.

Tol, R. S. J., & Dowlatabadi, H. (2001). Vector-borne Diseases, Development and Climate Change. *Integrated Assessment, 2,* 173–181. Retrieved 2-1-07, from http://www.uni-hamburg.de/Wiss/FB/15/Sustainability/iavector.pdf.

Tol, R. S. J., Ebie, K. L., & Yohe, G. W. (forthcoming). Infectious Disease, Development, and Climate Change: A Scenario Analysis. *Environment and Development Economics.*

Tol, R. S. J., & Yohe, G. W. (2006). A Review of the Stern Review. *World Economics, 7*(4), 233–250.

Toubkiss, J. (2006). *Costing MDG Target 10 on Water Supply and Sanitation: Comparative Analysis, Obstacles and Recommendations.* World

Water Council. Retrieved 8-1-07, from http://www.worldwatercoun
cil.org/index.php?id=32.

Townsend, M., & Harris, P. (2004, February 22). Now the Pentagon Tells
Bush: Climate Change Will Destroy Us. *Observer*. Retrieved
25-12-06, from http://observer.guardian.co.uk/international/story/
0,6903,1153513,00.html.

Travis, J. (2005). Hurricane Katrina—Scientists' Fears Come True as Hurricane Floods New Orleans. *Science, 309*(5741), 1656–1659.

Tubiello, F. N., Amthor, J. S., Boote, K. J., Donatelli, M., Easterling, W., Fischer, G., Gifford, R. M., Howden, M., Reilly, J., Rosenzweig, C. (2007).
Crop Response to Elevated CO_2 and World Food Supply: A Comment
on "Food for Thought . . ." by Long et Al., *Science* 312:1918–1921, 2006.
European Journal of Agronomy, 26 (3) April 2007, 215–223.

Turner, J., Lachlan-Cope, T., Colwell, S., & Marshall, G. J. (2005). A Positive
Trend in Western Antarctic Peninsula Precipitation over the Last
50 Years Reflecting Regional and Antarctic-wide Atmospheric Circulation Changes. *Annals of Glaciology, 41*, 85–91.

UN Millennium Project. (2005). *Investing in Development: A Practical
Plan to Achieve the Millennium Development Goals*. New York: United
Nations Development Programme. Retrieved 4-1-07, from http://
www.unmillenniumproject.org/reports/fullreport.htm.

UNCED. (1992). Rio Declaration on Environment and Development. United
Nations Conference on Environment and Development. Retrieved
30-1-07, from http://www.unep.org/Documents.multilingual/Default.
asp?DocumentID=78&ArticleID=1163.

UNDESA. (2006). *The Millennium Development Goals Report 2006*. New
York: United Nations Department of Economic and Social Affairs.
Retrieved 3-1-07, from http://mdgs.un.org/unsd/mdg/Resources/
Static/Products/Progress2006/MDGReport2006.pdf.

UNECE. (1996). *Long-term Historical Changes in the Forest Resource*.
United Nations Economic Commission for Europe and FAO, Timber
Section, Geneva.

UNEP. (2000). *Global Environment Outlook 2000*. London: Earthscan
Publications.

UNESCO. (2006). *Water—a Shared Responsibility: The United Nations
World Water Development Report 2*. New York: Berghahn Books.
Retrieved 7-1-07, from http://www.unesco.org/water/wwap/wwdr2/
table_contents.shtml.

UNFCCC. (1992). *United Nations Framework Convention on Climate
Change*. United Nations Framework Convention on Climate Change.
Retrieved 28-1-07, from http://unfccc.int/resource/docs/convkp/
conveng.pdf.

UNFCCC. (1997). *The Kyoto Protocol.* United Nations Framework Convention on Climate Change. Retrieved 18-11-06, from http://unfccc.int/kyoto_protocol/items/2830.php.

UNPD. (2006). *World Population Prospects: The 2004 Revision: Volume 3: Analytical Report.* United Nations Population Division. Retrieved 19-12-06, from http://www.un.org/esa/population/publications/WPP2004/WPP2004_Volume3.htm.

USCB. (1999). *Statistical Abstract of the United States 1999.* U.S. Census Bureau. Retrieved 9-12-06, from http://www.census.gov/prod/www/statistical-abstract-1995_2000.html.

USCB. (2006). Statistical Abstract of the United States: 2007. U.S. Census Bureau. Retrieved 30-1-07, from http://www.census.gov/prod/www/statistical-abstract.html.

USCB. (2007). Total Midyear Population for the World: 1950–2050. U.S. Census Bureau. Retrieved 2-01-07, from http://www.census.gov/ipc/www/worldpop.html.

USGS. (2005). Streamflow Trends in the United States. U.S. Geological Service Fact Sheet 2005–3017. Retrieved 22-12-06, from http://pubs.usgs.gov/fs/2005/3017/.

Utzinger, J., & Keiser, J. (2006). Urbanization and Tropical Health—Then and Now. *Annals of Tropical Medicine and Parasitology, 100*(5–6), 517–533.

van Lieshout, M., Kovats, R. S., Livermore, M. T. J., & Martens, P. (2004). Climate Change and Malaria: Analysis of the SRES Climate and Socio-economic Scenarios. *Global Environmental Change, 14*(1), 87–99.

Vandentorren, S., Suzan, F., Medina, S., Pascal, M., Maulpoix, A., Cohen, J. C., et al. (2004). Mortality in 13 French Cities during the August 2003 Heat Wave. *American Journal of Public Health, 94*(9), 1518–1520.

Varian, H. (2006, December 14). Recalculating the Costs of Global Climate Change. *The New York Times.*

Vaughan, D. G., Marshall, G. J., Connolley, W. M., King, J. C., & Mulvaney, R. (2001). Climate Change—Devil in the Detail. *Science, 293*(5536), 1777–1779.

Vaughan, D. G., Marshall, G. J., Connolley, W. M., Parkinson, C., Mulvaney, R., Hodgson, D. A., et al. (2003). Recent Rapid Regional Climate Warming on the Antarctic Peninsula. *Climatic Change, 60*(3), 243–274.

Vavrus, F. (2002). Making Distinctions: Privatisation and the (Un)educated Girl on Mount Kilimanjaro, Tanzania. *International Journal of Educational Development, 22*(5), 527–547.

Vavrus, S., Walsh, J. E., Chapman, W. L., & Portis, D. (2006). The Behavior of Extreme Cold Air Outbreaks under Greenhouse Warming. *International Journal of Climatology, 26*(9), 1133–1147.

Velicogna, I., & Wahr, J. (2006). Acceleration of Greenland Ice Mass Loss in Spring 2004. *Nature, 443*(7109), 329–331.

Vergano, D. (2006, August 3). High Heat: The Wave of the Future? *USA Today*.

Viguier, L. L., Babiker, M. H., & Reilly, J. M. (2003). The Costs of the Kyoto Protocol in the European Union. *Energy Policy, 31*(5), 459–481.

Vinther, B. M., Andersen, K. K., Jones, P. D., Briffa, K. R., & Cappelen, J. (2006a). Data for Extending Greenland Temperature Records into the Late Eighteenth Century. Retrieved 13-12-06, from http://www.cru.uea.ac.uk/cru/data/greenland/.

Vinther, B. M., Andersen, K. K., Jones, P. D., Briffa, K. R., & Cappelen, J. (2006b). Extending Greenland Temperature Records into the Late Eighteenth Century. *Journal of Geophysical Research–Atmospheres, 111*(D11).

von Storch, H., & Stehr, N. (2006). Anthropogenic Climate Change: A Reason for Concern since the 18th Century and Earlier. *Geografiska Annaler Series A–Physical Geography, 88A*(2), 107–113.

von Storch, H., Stehr, N., & Ungar, S. (2004). Sustainability and the Issue of Climate Change. Retrieved 26-1-07, from http://w3g.gkss.de/staff/storch/Media/climate.culture.041130.pdf.

Vose, R. S., Easterling, D. R., & Gleason, B. (2005). Maximum and Minimum Temperature Trends for the Globe: An Update Through 2004. *Geophysical Research Letters, 32*(23).

Walker, K. (2000). Cost-comparison of DDT and Alternative Insecticides for Malaria Control. *Medical and Veterinary Entomology, 14*(4), 345–354.

Wallström, M. (2001, July 2). European Climate Change Program: A Successful Approach to Combating Climate Change. *ECCP Conference*. Retrieved 6-11-06, from http://europa.eu.int/rapid/pressReleasesAction.do?reference=SPEECH/01/322&format=HTML&aged=0&language=EN&guiLanguage=en.

Waltham, T. (2002). Sinking Cities. *Geology Today, 18*(3), 95–100.

WCRF. (1997). *Food, Nutrition and the Prevention of Cancer: A Global Perspective*. Washington, D.C.: World Cancer Research Fund and American Institute for Cancer Research.

WDI. (2007). World Development Indicators Online. World Bank.

Weinberger, M., Oddone, E. Z., & Henderson, W. G. (1996). Does Increased Access to Primary Care Reduce Hospital Readmissions? *New England Journal of Medicine, 334*(22), 1441–1447.

Weisheimer, A., & Palmer, T. N. (2005). Changing Frequency of Occurrence of Extreme Seasonal Temperatures under Global Warming. *Geophysical Research Letters, 32*(20).

Weitzman, M. L. (2007). The Stern Review of the Economics of Climate

Change. *Journal of Economic Literature,* forthcoming. Retrieved 3-4-07, from http://www.economics.harvard.edu/faculty/Weitzman/papers/JELSternReport.pdf.

Wennberg, J. E., Fisher, E. S., Stukel, T. A., Skinner, J. S., Sharp, S. M., & Bronner, K. K. (2004). Use of Hospitals, Physician Visits, and Hospice Care during Last Six Months of Life among Cohorts Loyal to Highly Respected Hospitals in the United States. *British Medical Journal, 328*(7440), 607–610A.

Weyant, J. P. (1996). The IPCC Energy Assessment—Commentary. *Energy Policy, 24*(10–11), 1005–1008.

Weyant, J. P., & Hill, J. N. (1999). Introduction and Overview. The Costs of the Kyoto Protocol: A Multi-model Evaluation. *Energy Journal,* Kyoto special issue, vii–xliv.

WFS. (1996). *World Food Summit: Technical Background Documents, Docs. 1–15.* UN Food and Agricultural Organization. Retrieved 3-1-07, from http://www.fao.org/wfs/index_en.htm.

WHO. (2002). *The World Health Report 2002—Reducing Risk, Promoting Healthy Life.* World Health Organization. Retrieved 29-11-06, from http://www.who.int/whr/2002/en/index.html.

WHO. (2004a). *The World Health Report 2004—Changing History.* World Health Organization. Retrieved 13-11-06, from http://www.who.int/whr/2004/en/.

WHO. (2004b). *World Report on Road Traffic Injury Prevention.* World Health Organization. Retrieved 30-1-07, from http://www.who.int/world-health-day/2004/infomaterials/world_report/en/.

WHO & UNICEF. (2003). The Africa Malaria Report 2003. World Health Organization. Retrieved 29-12-06, from http://www.rollbackmalaria.org/amd2003/amr2003/pdf/amr2003.pdf.

WHO & UNICEF. (2005). World Malaria Report 2005. World Health Organization. Retrieved 29-12-06, from http://www.rollbackmalaria.org/wmr2005/.

WHO, WMO, & UNEP. (2003). *Climate Change and Human Health—Risks and Responses, Summary.* Geneva: World Health Organization.

Wiersma, A. P., & Renssen, H. (2006). Model-data Comparison for the 8.2 ka BP Event: Confirmation of a Forcing Mechanism by Catastrophic Drainage of Laurentide Lakes. *Quaternary Science Reviews, 25*(1–2), 63–88.

Wigley, T. M. L. (1998). The Kyoto Protocol: CO_2, CH_4 and Climate Implications. *Geophysical Research Letters, 25*(13), 2285–2288.

Wilby, R. (2004). Urban Heat Island and Air Quality of London, UK. Retrieved 17-11-06, from http://www.asp.ucar.edu/colloquium/2004/CH/presentations/AirQualityTutorialBackground.pdf.

Wilson, G. J. (1983). Distribution and Abundance of Antarctic and Sub-Antarctic Penguins: A Synthesis of Current Knowledge. SCAR and SCOR, Scott Polar Research Institute, BIOMASS Scientific Series no. 4.

Winfrey, O. (2006, December 5). A Green "Truth." *The Oprah Winfrey Show.* Retrieved 26-1-07, from http://www2.oprah.com/tows/past shows/200612/tows_past_20061205.jhtml.

Wingham, D., Shepherd, A., Muir, A., & Marshall, G. (2006). Mass Balance of the Antarctic Ice Sheet. *Philosophical Transactions of the Royal Society A: Mathematical, Physical and Engineering Sciences, 364*(1844), 1627–1635.

WMO. (2006, December 11). Press Release: Link between Climate Change and Tropical Cyclone Activity: More Research Necessary. World Meteorological Organization. Retrieved 18-12-06, from http://www.wmo.int/web/Press/PR_766_E.doc.

WMO-IWTC. (2006a). Statement on Tropical Cyclones and Climate Change. 6th International Workshop on Tropical Cyclones of the World Meteorological Organization. Retrieved 18-12-06, from http://www.wmo.ch/web/arep/press_releases/2006/iwtc_statement.pdf.

WMO-IWTC. (2006b). Summary Statement on Tropical Cyclones and Climate Change. 6th International Workshop on Tropical Cyclones of the World Meteorological Organization. Retrieved 18-12-06, from http://www.wmo.ch/web/arep/press_releases/2006/iwtc_summary.pdf.

Woehler, E. J., & Croxall, J. P. (1997). The Status and Trends of Antarctic and Sub-Antarctic Seabirds. *Marine Ornithology, 25,* 43–66.

Wood, R. A., Vellinga, M., & Thorpe, R. (2003). Global Warming and Thermohaline Circulation Stability. *Philosophical Transactions of the Royal Society of London Series A: Mathematical, Physical and Engineering Sciences, 361*(1810), 1961–1974.

World Bank. (2006). *World Development Report 2007.* Washington, D.C.: The World Bank Group.

World Water Council. (2000). *World Water Vision: Making Water Everybody's Business.* London: Earthscan Publications.

Worldwatch Institute. (2006). *Vital Signs 2006–2007.* New York: W. W. Norton.

Wunsch, C. (2002). What Is the Thermohaline Circulation? *Science, 298*(5596), 1179–1181.

Wunsch, C. (2004). Gulf Stream Safe If Wind Blows and Earth Turns. *Nature, 428*(6983), 601–601.

Wunsch, C. (2006). A Hot Topic. *The Economist.*

WWF. (2006, March 18). Canada's Western Hudson Bay Polar Bear Population in Decline: Climate Change to Blame. Retrieved 7-11-06, from http://www.panda.org/about_wwf/where_we_work/arctic/polar_bear/pbt_news_pubs/index.cfm?uNewsID=63980.

Xinhuanet. (2002, September 2). German Chancellor Urges All States to Ratify Kyoto Protocol. Retrieved 22-12-06, from http://news.xinhuanet.com/english/2002-09/02/content_547179.htm.

Yiou, P., Ribereau, P., Naveau, P., Nogaj, M., & Brazdil, R. (2006). Statistical Analysis of Floods in Bohemia (Czech Republic) since 1825. *Hydrological Sciences Journal–Journal des Sciences Hydrologiques, 51*(5), 930–945.

Yohe, G. (2006). Some Thoughts on the Damage Estimates Presented in the Stern Review—an Editorial. *The Integrated Assessment Journal, 6*(3), 65–72.

Yohe, G., & Neumann, J. (1997). Planning for Sea Level Rise and Shore Protection under Climate Uncertainty. *Climatic Change, 37*(1), 243–270.

Zachos, J., Pagani, M., Sloan, L., Thomas, E., & Billups, K. (2001). Trends, Rhythms, and Aberrations in Global Climate 65 Ma to Present. *Science, 292*(5517), 686–693.

Zhang, Z. X. (2000). Can China Afford to Commit Itself to an Emissions Cap? An Economic and Political Analysis. *Energy Economics, 22*(6), 587–614.

Zhao, H. X., & Moore, G. W. K. (2006). Reduction in Himalayan Snow Accumulation and Weakening of the Trade Winds over the Pacific since the 1840s. *Geophysical Research Letters, 33*(17).

Zwally, H. J., Giovinetto, M. B., Li, J., Cornejo, H. G., Beckley, M. A., Brenner, A. C., et al. (2005). Mass Changes of the Greenland and Antarctic Ice Sheets and Shelves and Contributions to Sea-level Rise: 1992–2002. *Journal of Glaciology, 51*(175), 509–527.

Index

Page numbers in *italics* refer to tables.
Page numbers beginning with 166 refer to notes.